DIGITAL MEDIA

PUBLISHING TECHNOLOGIES FOR THE

21ST CENTURY

FRANK J. ROMANO

College of Imaging Arts and Sciences
Rochester Institute of Technology

Micro Publishing Press
Torrance, CA

DIGITAL MEDIA

PUBLISHING TECHNOLOGIES FOR THE 21ST CENTURY

Frank J. Romano

Published by:
Micro Publishing Press
2340 Plaza Del Amo, Suite 100
Torrance, CA 90501
(310) 212-5802

First Printing: July, 1996

Printed in the United States of America

ISBN 0-941845-18-4

FOREWORD

Johann Gutenberg (that was not even the real name) did not invent printing. The concept and the term existed in his time, and even before his time. Wooden blocks were carved with the text and images in relief and had been used for reproduction for many years. He invented moveable type.

Gutenberg was affiliated with the goldsmiths' guild and was probably raised with the secrets of this craft—carving, polishing, working with metals. For a while he lived at a monastery, which, like most at that time, had a scriptorium, where monks religiously scribbled their way to Heaven. He must have observed the laborious, time-consuming handwritten production of books.

By dividing his language into its modular units he was able to design type that looked like the many variants used by the scribes. His font eventually contained 292 pieces of metal, many of them 2-letter and 3-letter combinations of connecting letters. There were even six versions of the lowercase i. His secrecy was based on the mass production of printed books to be sold as handwritten books.

It took him three attempts and almost a decade to get the font right. There was a 36-line, a 40-line and a 42-line version (based on the number of lines to the Bible column). Later on, his partner, Fust, took some Bibles to Paris to sell. The buyers flocked and paid "handwritten" prices. They compared and found the same errors on the same pages. They were exact. This had to be the work of the devil.

Actually, the first publication that came off the Gutenbergian press was not as lofty as the Bible. The first printing was an Indulgence, which was a sort of passport to Heaven. It had blank areas for a name and an amount to be filled in. Yes, the first printing from moveable type was a form!

Within fifty years of the invention of moveable type, the scribes were virtually out of business. The printed book was well established and printers like Aldus Manutius helped to create the Renaissance by putting in print the works of Greek and Roman philosophers, mathematicians, historians and others. Printers like Caxton gave us the English language by "freezing" in print the communication agglomeration that evolved from the Britons and the Norman French, to say nothing of the Romans and a few barbarians.

These early printers distributed information and knowledge. They were the manufacturers and the marketers. They moved more than type. They moved the very substance of our culture: *that which we know.* They made it possible for us to know more. They provided the tools for learning—units of information and knowledge called books.

The concept of the printed page which made a book also made a newspaper and a magazine and even junk mail. It was print that created Western civilization. In addition to its universality of delivery it also became a reference, a repository for knowledge.

The only demand that print placed upon the user was the need for literacy. The act of reading forced some level of thinking. Will new electronic media reinforce our cognitive skills or will we become passive observers of vicarious occurrences on computer screens?

This book was created to put in one place at one time an overview of the challenges facing print and knowledge transfer in the last few years of the 20th century. Here they are in all their glory.

A special thanks to Eric Cohen, my indispensible grad assistant, for the HTML chapter. I really miss him now that he has graduated and is working in the real world. Thanks to Magda Knaflewska for proofing the manuscript.

And thanks to the many students in my *Electronic Publishing* classes who struggled with me to understand the new approaches to information presentation and delivery, and the second revolution in moveable type.

Frank J. Romano
Salem, NH

TABLE OF CONTENTS

Foreword v

1. The Information Society 1

2. Print Media 15

3. New Media 29

4. Interactive Multimedia 37

5. Portable Documents 43

6. SGML and HTML 71

7. The Internet & The Web 115

8. CD-ROM & Digital Video Disk 133

9. Intellectual Property 139

10. Telecommunications 159

11. The Future 167

Afterword 179

Index 183

About 40 years after the invention of moveable type, the Abbot of Sponheim, Germany published an impassioned book called *In Praise of Scribes,* extolling the glory of handwritten books. He argued that scribes should continue to copy manuscripts despite the advent of printing because of the need for diligence and devotion exercised in the process. He also said that hand-copied words on parchment would last centuries beyond printed words on paper. To get his message out, the good Abbot had the book printed. He used new technology to promote old technology. The Abbot of Sponheim is a metaphor for contradiction. Just like:

- The postmasters of America using e-mail.
- The Postal Service Web site.
- The parts catalog for a Heidelberg press on CD-ROM.
- Disks for AOL & Compuserve in magazines.
- Print ads for World Wide Web sites.
- Radio spots for Miracle Ear.
- Books on how to read.
- *Wired* magazine in print.

Today, we live in a world of contradiction, in a nation that produces smart bombs and dumb kids—hesitant to embrace new technology at the expense of old technology.

1.

THE INFORMATION SOCIETY

It is truly a brave new world. I received e-mail from 80 or so people who responded to an article I wrote about the 20th anniversary of a publication I founded. My first reaction was: who ever dreamed we would communicate this way back in 1978? How our world has changed. Ponder this:

1. 40% of U.S. households control 70% of the disposable income. These tend to be affluent, educated Americans with access to the latest technology and the latest information.

2. Of total personal consumption, 6.7% goes for movies, theater and sports, 2.2% goes for magazines, .8% for newspapers, and 1.9% for books. 62% of all adults read a newspaper daily and 75% of all adults read a newspaper at least once a week. U.S. adults read or scan an average of 10 magazines a month. We *watch* more than we *read*.

3. Most of us have only about 3,500 hours per year of personal time that can be devoted to reading, TV, listening to music, etc. That includes weekends, holidays and vacation, plus whatever you can grab during the work week. Thus new media compete with old media for our attention.

4. Cost per thousand baud telecommunications was $500 in 1980 and is about $22 in 1996. Eventually it may be cheaper to transmit information than to record or print and deliver it to you. CD-ROM could yield to online delivery over time.

5. In 1986 every encyclopedia published was only in print. Today, 27% are in print, 33% on CD-ROM, and 40% on a CD-ROM bundled with your new personal computer. The encyclopedia is essentially free for some people. Eventually 90% of all encyclopedias could be digital. Watch for the encyclopedia to go on the Internet once a business model can be worked out.

6. People who work at or from home for any amount of time represented 28% of the workforce in 1988; in 1995 39%. It could hit 50% by 2000. Not all are home offices as such.

7. Students per computer in 1984 averaged 125; today it is under 9. More people are entering the workforce with computer skills. Eventually, they will not be able to enter the workforce without computer skills.

8. The average number of working hours per week in 1964 was 38; today it is 52! There are more 2-person wage earners in a family and the home office is an extension of the real office. We work harder. We need to work smarter.

9. U.S. movie ticket revenues were $6 billion in 1995. 1 out of 4 U.S. adults goes to the movies once a month. But they rent one video cassette a week as well. U.S. bookstore sales: $11 billion, 24% of which is sold in two summer months. I guess you still can't read the computer out of doors.

10. Average daily household television ***on*** time is 6 hours and 39 minutes. This number is actually rising and represents 2,427 hours out of the 3,500 hours of personal time available. We actually watch the TV an average of 4 hours a day. On weekdays, adults listen to the radio for an average of 3 hours and 20 minutes, in many cases, in their automobiles.

11. One in ten households has a desktop computer with a modem. In 1995 the number of e-mail messages exceeded the number of print mail messages transmitted in the U.S.

12. In only 7% of U.S. households is the TV only for broadcast reception. In most, the TV is connected to cable, video cassette recorder/player and/or a video game system. It is not too difficult to predict that the TV will become a window to the Internet.

13. 72% of American teens have a CD player. Each of us spends $55 a year on recorded music. Actually my kids more than cover their share.

14. 31% of American teens have a home computer. Just because it is in the home does not mean it is available for business use. Both kids and parents will compete for the computer . . . and the TV.

15. Average life expectancy for men today is 63; in 2010: 67. For women, 67 and 71. If we develop a customer they may be around for a long time.

16. Worldwide population of personal computers is now 150 million; by 2010 over 300 million. Most are too old to do what we want to do today.

17. Number of McDonald's fast food restaurants: 14,000; by 2010: 30,000. We want what we want when we want it. The same will be true of information.

18. Gambling revenues in 1994 were $39.5 billion. By 2010 it is expected to be $126.6 billion. Wanna bet?

19. There are 1,100 communications satellites. In 2004 there will be over 2,000. Everyone will be able to connect.

20. Of all the people ever born on the planet Earth, 51% are now alive. That's why you can't find a parking space when you really need it.

The virtual world

It is clear that there is competition for our time and attention. The virtual communities talked about during the 90s have become a common part of both our business and social lives. The capabilities of the TV, PC and the telephone have merged into a single type of "Information Node" that provides an interface to the outside world. This machine, now called a computer, has become a standard part of lifestyles at the end of this century; in 2003, 80% of all homes will probably have one.

This addition to consumer communication and entertainment choices is an extension of the trend observed regarding several media technologies in the late 80s and early 90s. From telecommuting to home banking, we can connect into another world—a virtual world—also known as cyberspace.

An information rich society

We are an information rich society and in our private or business lives we seek out information, process it and apply it for a multiplicity of reasons. No one has yet tracked the number of pieces of paper we come into contact with over a year but it is as large as the number of electronic "pieces." The number of competing messages is very large and marketers of the future will have many channels for reaching potential customers. The challenge will be getting a message read among all the messages that will bombard us.

A technology-rich society

There are 250 million Americans. Traditionally there have been two places where Americans work: at their business or in their home. Now there is a third place where they work: on the road. In the home, we know:

- 96% of U.S. homes have telephones
- 59% of U.S. homes have cordless telephones
- 21% of U.S. homes have cellular telephones
- 60% of U.S. homes have answering machines
- 98% of U.S. homes have television sets
- 88% of U.S. homes have VCRs
- 62% of U.S. homes have cable television
- 16% of U.S. homes have modems
- 46% of U.S. homes have videogames
- 48% of U.S. homes have CD audio players
- 40% of U.S. homes have PCs
- 19% of U.S. homes have PCs with CD-ROM
- 34% of U.S. homes have computer printers
- 8% of U.S. homes have fax machines

We are told that the computer will re-shape our spare time and our business time. Thus, there will be information *haves* and *havenots*. The concept has been that homes will have to have computers to connect to the Internet. However, cable modems and on-set converters will transform the TV for de facto computer access. At that point TV may be less of a business oriented system and more of an entertainment system. When a device can be shared by other family members, it will yield to shared applications.

Offices in the home

54% of American homes have an office of some kind—53.1 million households. In many cases it is used to balance the checkbook and perform family finances. The estimated number of home businesses is 4 million. The *home office* is used occasionally and the *home business* is used frequently.

The differences in technology use are minimal. The home office worker sees their home office as an extension of their real office and attempts to emulate the technology that they are accustomed to. The home business user sees their home office as a business location. Of all home offices:

- Over half have a computer

Software in use:	
Word processing	90%
Spreadsheet/accounting	67%
Communication and e-mail	30%

- An average of 35% have a fax machine
- An average of 40% have fax modems
- An average of 55% have printers
  A majority print more than 40 pages per month.
- An average of 42% have CD-ROM drives
- About 19% are connected to an online service
- About 15% have a copier
- About 9% have a scanner of any type

As one can see, both kinds of home office workers tend to be more technology savvy. Both home office user types are major customers of convenience reprographic services. As the price drops and the capability increases, more home office users will acquire technology that lets them:

- Create text and presentation based documents
- Use more graphics and color
- Print them out at 600 dpi black
- Connect to the Internet or other online services
- Use e-mail routinely
- Access the World Wide Web for information
- Travel on business with a portable computer
- Use a cellular telephone or a cellular computer

The citizen of the future will be more mobile, more technology oriented and have needs that are more sophisticated. U.S. Internet user profile (1995 survey)

Uses	
Browsing	83%
Entertainment	57%
Work	51%
Education	47%
Business research	38%

Corporate usage of the Internet	*1995*	*1996*	*1998-2000*
Product transactions	14%	34%	44%
Customer service Tech supt.	51%	66%	73%
Communication w/suppliers	62%	73%	78%
Document publishing	56%	68%	71%
Competitor research	56%	65%	67%
Job posting/Search	60%	66%	69%
General research	19%	19%	19%
Playing games	27%	27%	27%

Technology changes

Computers

Measured in Mips (millions of instructions per second) per dollar, computing capability doubles every 18 months. More and more computers are portables with integrated capabilities for CD-ROM, modem connections and color screens. Moore's Law says that computer power increases every 18 months.

Scanners

Low cost scanners are proliferating. Monochrome scanners are as cheap as $99 for hand-held units. Good quality color scanners are under $1,000 and should achieve prices at $500 by 1998. There is no doubt that the scanner will be a common peripheral for computer users.

CD-ROM

The next generation of CD-ROMs will hold 4.7 gigabytes, up from 650 megabytes today. That is sufficient for just under 2 hours of high resolution video or hours of sound or hundreds of thousands of pages. Color image and document storage will be easier.

Personal CD-ROM writers have dropped from over $10,000 to under $2,000. Soon they will be $500 and every computer user will be able to make their own CDs.

Modem

In less than a decade, modem speeds have jumped from 300 baud to 28,800 baud and beyond. Half of what passes over the U.S. telephone networks is data—and it represents 20% of the profit. The telephone network is becoming a computer network. Thus computer users will be able to create and have WWW sites maintained by service providers.

Color printers

In 1989 the first thermal desktop color printers were introduced at $20,000 and dye sub color printers were $75,000. Today, color printers average $500 with the best of them at no more than $10,000. Eventually many users will have access to color printers for low volume requirements.

The American home and small business will be equipped with the technology to create text and graphics documents and presentations, make CDs, connect to the Internet, print out pages in color and even provide material for others to access electronically.

What it all means

We will become information providers as well as information consumers. That sums up the crux of what is going to occur. Even the smallest business will have a presence on the WWW. They will need a service to help them create the site and maintain it. The user will not want to maintain their own server and reprographics services will offer that service.

But, is it paper?

The cost of paper-based publishing is rising. Print is challenged by alternative media because:

- Paper costs are rising, probably never to return to their historical levels.
- Distribution costs (including postage) are based on labor and energy costs and can never get cheaper.
- Out-of-date information places a burden on business and industry as well as consumers.

The expectations of the information consumer are also rising. As literate consumers get their hands on a mouse and use a computer, a revolution of rising expectations will sweep publishing markets. We can expect:

- Faster print turnaround and higher quality to compete with electronic media.

• Searchability of interactive products to go beyond the the passivity of paper.

• Pay-per-view for information that is needed rather than information that may never be used.

Publishers now fall into four broad classes:

• Commercial publishers print for profit and produce newspapers, magazines, journals, books and more.

• Corporate publishers create documents for advertising, marketing, corporate communications, technical documentation and collateral material.

• Government publishers generate landfills of legislative, regulatory and judicial documents.

• Personal publishers are Web users who create their own home pages. The truly disadvantaged in our society will be the homepageless.

Who would have thought that each of us would become our own publisher.

Digital media

Publishers now fall into four broad classes:

• Commercial publishers print for profit and produce newspapers, magazines, journals, books and more.

• Corporate publishers create documents for advertising, marketing, corporate communications, technical documentation and collateral material.

• Government publishers generate landfills of legislative, regulatory and judicial documents.

• Personal publishers are World Wide Web users who create their own home pages. The truly disadvantaged will be the homepageless.

The new world

The world of electronic publishing has opened entirely new market places to publishers. Printed books may one day be just a distant memory, delegated to being only museum pieces. The computer age has quickly descended upon us and is becoming an everyday part of life for nearly everyone in the world. The "Information Superhighway" has shown great promise in distributing information to large numbers of people. CD-ROMs have given us the ability to store vast amounts of information and entertainment in one place, and even be interactive.

The standardization that occurred in the network infrastructure has been matched by business hardware and software vendors. While differences in hardware architecture still exist, application program interfaces and software development tools buffer application developers from these differences and allow coordinated development across multiple platforms. Also, robust application development tools have simplified application development. Windows NT has become the operating system of choice, as UNIX has been unable to keep up the changes to support the new applications.

This new environment has affected the application development team organization. The programming position of these teams has disappeared as technical skills have diminished in importance and business skills have become the key to successful application development. Thus, application development costs have plummeted and a much wider range of companies customize "off-the-shelf" software. Systems integration and software companies offer object libraries at reasonable prices that allow easy integration of multimedia features to business applications.

The concept of what we now think of as a book may someday be completely rewritten. Technology could give us computers small enough to literally fit in our pockets, with enough battery power to last days, weeks, or possibly, even years. Imagine, something so small as a checkbook, being able to access the world through wireless technology such as satellites, microwave transmissions, or maybe something entirely different. The system might have voice recognition instead of a keyboard, and have high definition 3-D graphics for interactive pleasure.

Classroom textbooks could soon be replaced with any number of technologies available to us today. CD-ROMs can hold volumes of information and be accessed quickly and easily through search and retrieve functions. Short instructional videos can also be included for lab manuals, or history books could have clips of some of the world's great leaders delivering the most famous of their speeches.

The possibility to create custom books is also present. Instructors are using custom textbooks made up of chapters of a number of texts. Magazine publishers are compiling articles to make unique publications. They could also be digital.

Children's books will also become more enjoyable for the child, as well as being more educational. Many children now may be turned off by the passiveness of reading books, but an interactive CD-ROM would allow the child to be in control and read what they want to. Basic techniques of using a computer will also be taught while the child both enjoys and participates in their story.

Resource books are another area that electronic publishing will eventually dominate. The ease of using and locating the information you want makes CD-ROMs extremely versatile in this area. Encyclopedias, dictionaries, and indexes could be continuously updated through the Internet. Another advantage of using CD-ROM for such an application would be the great amount of physical space that is saved over printed copies. A complete set of encyclopedias can be stored on several CDs instead of an entire shelf. A fully stocked library of information could be stored on your desktop.

But the idea of books in hard copy will never totally disappear. A computer will never be able to replace decorative books like the ever popular coffee table versions. While computers are becoming more "user friendly" as time progresses, they will never possess the "warm" feeling of printed books. I have yet to hear anyone say they want to curl up in bed, or in front of the fireplace, with a good laptop computer. A drawback to electronic publishing is that not everybody has access to computers in order to take advantage of the many positives of the technology.

In the strictest definition of the word, the printed, bound volume, the book, is going nowhere, so to speak. The book is the most portable of all forms of communication. It requires no form of outside power in order to be read, it is easily accessible, and can be used anywhere. By anyone.

In the words of Clarence Day, "The world of books is the most remarkable creation of man. Nothing else that he builds ever lasts. Monuments fall; nations perish; civilizations grow old and die out; and, after an era of darkness, new races build others. But in the world of books are volumes that have seen this happen again and again, and yet live on, still young, still as fresh as the day they were written, still telling men's hearts of the hearts of men centuries dead."

Digital dominance

Our government has supervised the merger of the telecommunications, computing, and entertainment industries. In developed nations around the world, governments made strategic "information highway" investments, led the formation of networking standards, and most importantly, set the rules for competitive play in the marketplace. The mergers of the 90s forced government to quickly re-evaluate their existing telecommunications regulations to provide a level playing field. The Modified Final Judgment which broke AT&T into the 7 Regional Bell Operating Companies (RBOCs) in 1984 was overhauled. This created a level playing field for network providers allowing telephone companies (both local and long distance carriers) and cable companies to compete fairly in all aspects of providing phone and television services to the consumer. The new Telecommunications Act frees AT&T to re-enter the market and gives the RBOCs greater opportunities for expansion. These changes in the United States are expanding across the globe, leading to the worldwide infostructure being controlled by a small number of loosely regulated mega-corporations.

Services offered by the large information provider/computer service companies to support businesses include communication services (wire and wireless telephony, video-conferencing, and e-mail/voice-mail/video-mail services), information services, business process management and re-engineering services, systems integration services for wide area networks, packages for common business transaction and logistics management systems, and CAD/CAM applications.

Congress has required the Federal Communications Commission (FCC) to oversee the final steps of the process. Information Carriers, as they came to be known, would act as common carriers. Parent companies might, through other subsidiaries, also supply content; but all the Information Carriers would be subject to the same terms, free to provide network access of all types (telephone, television, and data) to businesses and consumers. A usage-based fee logic was developed that compensated the Information Carriers by a combination of time charges and transaction fees. State Public Utilities Commissions were strong-armed into consenting to these same measures. In return, competition was allowed at every geographic level and incentives were provided for carriers to serve all parts of the world.

Despite providing a competitive business environment for the worldwide telecommunications providers, the U.S. government was unable to deal successfully on other major issues. For over a decade, Court rulings steadfastly refused to address the needs for simplified intellectual property, copyright legislation, and data security.

Information will be like electricity. In your home, office, car, on a train or plane, almost anything electronic can be plugged into the world of data, once considered and named the *information superhighway*. Children are shown the world by dealing with in a virtual manner, and countless of their parents are employed in creating it at work. Communities appear where people congregate to work, shop, discuss, and entertain themselves, all electronically and facilitated by interactive media technology. We don't even see the difference between the TV and the computer and the telephone as they create profound changes in both the consumer and business worlds. The power of the network allows small companies to provide the services that, in the early 90s, would have required a large organization.

In this networked environment, interactive software has flourished. It can be transmitted over the Net with high-definition, 3-D pictures and holographic sound. Also, because of the two-way features of the Net, practically anyone can be a content provider by feeding applications back into the network from a properly configured computer system. Of course, a key to being a successful content provider is to create interesting and interactive applications, a skill that is difficult to master. For those that do, however, there is strong demand for both entertainment and education services.

In the early 1980s, William Gibson wrote a science fiction novel. *Neuromancer* involved large corporations that replaced governments, and freedom-fighting computer hackers who waged war against secure data. The novel set the stage for a genre of science fiction that would have a profound influence on the computer and telecommunications community. A large part of the plot unfolded in a setting that had no physical existence—called cyberspace—a consensual hallucination that felt and looked like a physical space but actually was a computer-generated construct. People could plug into data systems and networks and have the sense they were actually entering a place that had no correlation in physical reality. People carried out business transactions, communicated with one another, worked, played, and, as they have done in every other place they had occupied, broke the law. Gibson's vision saw the ever-increasing mass computer connectivity as more and more people and systems were going online. The term cyberspace caught on as a matrix of interconnected computers. Millions of Americans are interacting every day in a world they call cyberspace.

The mechanics of this interaction involve a user with a personal computer who makes a connection with another computer through telephone lines. The user's computer and the remote host system become linked to create the perception that there is no physical distance between them. When computers communicate there can be immense amounts of data transferred. This entire book can be sent across the globe in just a few nanoseconds. It could also be financial data, digitized audio and video, and computer software. Instead of waiting days for a letter to make its way through the "snail mail" postal system, a user can have electronic mail delivered instantly to distant locations. Users who are connected to the same system at the same time can "chat" in real-time. Several users can interact in the computer equivalent of a conference call. This new medium of communication brings with it major challenges.

Electronic publishing

To become a publisher, you need a printing press and a method of distribution so that the printed word can reach its audience. With the advent of cyberspace, the nature of publishing changed dramatically. Online systems and networks make distribution of electronic publications feasible even to individuals with limited resources. There are now countless electronic periodicals (often called *zines*) distributed to cyberspace cadets. One can subscribe to these periodicals by sending e-mail over a network to the publisher so that each issue is sent directly to the subscriber's mailbox. *Zines* are frequently archived by BBSs and online services so that every back issue is available upon demand. Such systems are similar to traditional newsstands where a multitude of titles is made available to the public. Usually, no fee is attached to acquiring zines, although commercial zines are becoming more prevalent. Besides the zines, which usually are not found in paper format at all, there are many electronic versions of conventional paper periodicals. These are sometimes distributed at the same time or even before their paper counterparts hit the newsstands. Most major newspapers and magazines can be found in digital form in cyberspace.

As consumers and business have integrated technology; as government and industry have created the technical infrastructure; and as publishers, both commercial and private, have developed content, we have moved inexorably towards a new information society.

2.

PRINT MEDIA

What has a gigabyte of immediately accessible and randomly accessed memory, a user friendly high resolution graphical user display and interface, and the ultimate in portability? It could be the latest computer laptop or palmtop, but more likely it's the latest issue of a magazine or newspaper or even a catalog.

You have heard that print will change over the next decade or so. Actually, print will not change—the information that was disseminated by print will change. Print may be produced with different or improved technology, but it will still be print on paper and that paper will have to be mailed or delivered to you.

At the same time there is a growing group of electronic competitors to print as a vehicle for information delivery. Catalogs and reference works are now routinely released on computer disks. They display pages in color, with graphics and sound and even video. It is called multimedia because it combines visual and audio information in new and exciting ways. The only limitation is that you need a computer.

Disks can be floppies, CD-ROMs, flopticals, removables and non-removables of all kinds. There is no standardization and someday there may be data archeologists whose only job is to identify strange electronic storage media from the last decade. Don't believe me? Go into your basement and identify the old 78 rpm turntable, 8-track tape cartridge and Phillips cassette player, which have been replaced by your new compact disc player or DAT tape player upstairs. Will you be able to play the Beta video tapes of your wedding in the year 2034? Recorded media keep changing with smaller units that hold more data. In contrast, the other day I read a book that was printed in the year 1501.

Data highways and byways

The great data highway will have us sitting at computers connecting into online services and universal e-mail or voice mail (which I guess is v-mail). Think of the data highway today as CB radio, only you have to type.

Broadcast media will give us 500 channels of cable TV with downloaded movies and entire channels devoted to "Gilligan's Island." The telephone companies and the cable companies are fighting for the link into your home. It is your on ramp to the data highway and there will probably be a toll booth. You will have a computer, whether you want it or not. That remote control "clicker" that you now use to surf through 80 channels will activate a new control box that will do what a computer does. Your Yellow Pages will be electronic and so will your shopping.

Hewlett-Packard has a color printer that hooks to your TV so you can print screenfulls of information on paper. Instead of printing one million pages on a printing press, one million users will print one page on their home or business desktop or production printer.

Thus print as we know it is under attack. New media proponents bombard us with statistics: half of all magazines displayed at the newsstand are thrown away because they are not sold; 40% of what goes into landfills is paper. A book only needs to sell 100,000 copies to be a bestseller, but 42 million people watch "Roseanne."

So they say: let's make magazines electronic. You can download only the information you want or need. Catalogs could also go that route with multimedia presentations of products. But in both cases there still is print. Your electronic magazine will find some of its information on paper. Not from a large printing press, but from your office or home printer. The electronic catalog will let you find what you want but you will print the page out for reference.

Let's be honest: are we all going to carry around computers? Are we going to spend our days at the office in front of a computer and our time at home huddled around the TV screen? Will every person on the 7:42 commuter train be reading their computer? There is that scene in "The Graduate" where Dustin Hoffman is told to invest in plastics.

I say: invest in batteries. The tools that we will use to do our work will be electronic and they will require energy.

Telling you how wonderful print is will not change the inevitability of change. Our entire culture is derived from the printed word. Our powers of concentration are based on the impedance factor between the eye/brain and the printed paper. The ephemeral images of the video screen change all that. There is to come the Nintendo Generation with the attention span of non-vertebrates. But we are powerless to stop the rush of technology.

There are benefits to new technology. In my class on "Electronic Publishing" I throw one day of the "Congressional Record" (some 500 to 900 pages) and ask the poor unfortunate student who happened to sit in the front row to find every occurrence of Senator Kennedy speaking in the well of the Senate about an amendment to a bill on comparable worth. The student goes through it for a while and gives up. What were they looking for? After a while they say it: An index. So I throw down the every-two weeks cumulative index (about 400 pages). Call up the CD-ROM or Web site, search to retrieve the information on screen, and print.

By law, there are depository libraries in the United States that must get the Congressional Record. Some now get the print version and some get the microfilm version. The electronic version could eliminate all printing of the Record, but not all of these libraries have computers so they continue to accumulate the printed Records and at some point in time, discard them. Thus, the Government Printing Office will still be printing the Record long after the electronic version has taken hold, but in smaller and smaller quantities.

That brings us to the concept of demand printing. The traditional printing press is geared for longer runs of large sheets (called signatures). It has not been cost effective to handle short runs—those under 5,000. This is changing as new plate and press technology are adopted. At the same time, digital printers are evolving for runs well under 5,000. This will give rise to new forms of publishing. In academia it is now possible to have a course book customized with chapters from a number of books. McGraw-Hill has been a pioneer in this area.

The concept of "just in time" delivery is based on printing what you need when you need it. Instead of producing a million units of a promotional brochure with information that may change and require many of them to be discarded, the JIT philosophy says that you produce in those quantities required at the time. This will also affect traditional print run lengths. The ultimate in on demand publishing is the advent of the fax newsletter. The information magically appears in print from your fax machine.

Print is fighting back. New plate technology allows offset lithographic printing without water (just ink) which is capable of 400 line screen. Stochastic (or frequency modulated) screening eliminates halftone dots as we know them to provide high levels of print quality that have never been achieved before. New output devices generate multi-page imposed plates directly from electronic data. New presses automatically load plates and cut makeready times. Other presses make plates right on the press from digital data. High fidelity color printing is aiming to achieve the quality level of photographs. There is a link between traditional photography and print with Photo CD, and there are technologies that are moving print toward continuous tone reproduction.

They call all of this "high end printing" to indicate that print is changing for the better. It is adapting in order to lower cost and increase quality. Even if you utilize electronic approaches, print is handy for a number of reasons. You can make comments in the margins and emphasize information. You can pass it to almost anyone. Electronic approaches require computers, and we do not see our entire society computerized.

Then there is the book. The book that is read for pleasure: the mystery, the romance, the western, etc. It is portable, easy to use, requires no energy, and can be re-cycled to other readers, none of whom need a machine to read it. Print is not dying; print is changing and it will be around for a long time.

Paper is the culprit

Back in 1995 paper prices went through the roof. Most printers were paying 30% or more for certain grades. So what was the difference a year made?

Actually our story started a few years before. The economy was sluggish and commercial and inplant printers were not doing very well. Paper was plentiful and you could almost name your own price. The pulp producers then made the decision to sell pulp overseas at excellent rates, and these contracts were for three and four years.

Exit pulp. The economy gets better and so do commercial printers. If you do not know it already, the printing industry tracks the GNP almost point for point. Printers then needed more paper. Then, still more paper. Suddenly demand outstripped supply. Economics 101: prices go up for the available supply. Large users went on allocation which means they may only get the amount of paper they purchased the previous year. Certain grades were hard to come by. It affected different printers differently, but it will affect all of us one way or another.

It is so bad that counterfeiters are bleaching the green out of currency since the paper is more valuable.

On demand to the rescue

With over 400 digital presses in the U.S. soon and lots more coming, short run, on demand approaches could be arriving just in the nick of time. Print buyers can now purchase only what they need. They may not have as much selection in terms of paper grades, but most of what will be available will be more than acceptable.

There is a major business opportunity here. Digital printers can hold documents and images in electronic archives and print them as needed. Instead of keeping a big room full of film flats (negatives) the printer will have a big storage system and great retrieval capabilities to print on demand—anywhere. For a centralized operation like Donnelley's Digital Division the material could be FedExed (FedEx is now a verb) to customers, or for operations that may have facilities in multiple cities, like Moore's stated plans, the files could be downloaded to the regional facility for printing.

The beginnings of demand color reproduction

In 1938, the first electrophotographic image was produced by Chester Carlson, a physicist and patent attorney.

Financial and market success did not come immediately. It was not until 1944 that Carlson was able to persuade Battelle Memorial Institute to conduct additional experiments with his process. During the period from 1944 to 1948, the basic process and technology were refined. From 1944 to 1946 Battelle invested its own resources, but in 1946 the picture changed. The Haloid Company (which eventually became Xerox Corporation) of Rochester, NY, was granted a license to develop electrophotography, and later, in mid-1948, the U.S. Army Signal Corps sponsored research at both Battelle and Xerox, intending to apply the process to military photographic areas. The Xerox Corporation's first commercial product based on electrophotography (renamed xerography, from the Greek words for dry writing) was a hand-operated copying system which appeared in 1950.

In 1960, the first automatic office copier—the Xerox 914—appeared on the market. At six copies per minute, the 914 had most of the capabilities of today's "personal copiers." In 1970, IBM marketed the Copier I which featured an innovative, reusable, organic photoreceptor. Five years later, the Eastman Kodak Company, a company which already had a century of experience in the imaging field, joined the electrophotographic copier industry with its Ektaprint series. The latter system incorporated a re-circulating feeder for originals and an online binder/stapler. It was the beginning of demand printing.

Many United States companies developed remarkable products to take advantage of growth in the reprographics market. All but a few have abandoned their efforts in the face of Japanese competition.

Color electrophotography

Color electrophotography began with 3M's Color-in-Color copier in 1969 which used electrophotography to form a mask that controlled the migration of colorants. In 1973 Xerox released the 6500 Color Copier, and Canon followed in 1978 with the Canon T machine.

Eastman Kodak entered the color copier arena in 1988 with the ColorEdge copier. At 23 three-color copies per minute and 70 cpm for single color copies, the ColorEdge represented technological innovation in electrophotography. The ColorEdge used a flexible photoreceptor belt architecture. The original was placed on the platen and a flash exposure through each of three filters—red, green, and blue—performs the color separation. Development of the electrostatic image was via a magnetic brush. Kodak used the transfer directly-to-paper technique to synthesize the image from the three colorants. The last toner transferred is yellow, and then the image is forwarded to the heated roll fuser. The ColorEdge could not copy in black only.

The Xerox 1005 is the oldest of the full color copiers considered here. Its genesis is the Fuji-Xerox 6800, a second-generation of the original Xerox 6500. The 1005 was a drum copier with a scanning optical system that performed the color separation function via the three filter exposure technique. It had the usual three-development stations around the periphery of the photoreceptor drum. Transfer was directly to the paper on an intermediate transfer roller.

Other color printers

In 1987, Mead Imaging announced its new color imaging technology Cycolor. Cycolor technology represented nearly a decade of development at the Mead Corporation and its subsidiary, Mead Imaging. The Cycolor system was the first practical non-silver halide, dry process, high resolution, continuous tone color imaging technology. It was based on microencapsulation, the same technology used in carbonless paper. Cycolor was a two component imaging system consisting of the film or donor sheet and the receiver sheet. The Cycolor film is coated with billions of cyliths each only several micrometers in diameter. First, the film is exposed to the colored light. Those cyliths sensitive to the colored light become hardened, while the unexposed cyliths remain soft. The donor is then pressed against the receiver sheet by the pressure rollers. The soft cyliths burst under pressure, transferring their dye to the receiver. The dye precursor from the burst cyliths then reacts with chemicals on the receiver sheet, forming the appropriate color. Continuous tone capability was possible, since cyliths harden in proportion to the amount of the light exposure. It all never really worked.

In terms of color printers, in 1985 Howtek demonstrated the HR-1 "Thermo-Jet" solid ink jet printer. Exxon and Dataproducts announced an agreement to cooperate in the commercialization of Exxon's solid ink jet technology named E/D Venture, which resulted in a monochrome office printer called the SI (for Solid Ink) 480 by Dataproducts in 1986. Howtek continued to develop its color solid ink jet printer renamed the Pixelmaster.

The first thermal dye transfer printer was the the Sony Mavigraph and was shown as a prototype in 1982, the same year as Shinko Electric and Seiko introduced the first conventional wax thermal transfer printers. The Mavigraph did not become commercially available until the spring of 1986.

Only within the last decade has color printing technology developed the methods and processes to create documents at the quality, speed and cost that users can justify.

The roots of digital demand printing

In 1977 Xerox introduced the 9700 laser printer. It ushered in an age of digital laser printing that eventually found its way to virtually every desktop. It also spawned considerable interest in, and eventual application of, the concept of demand printing and publishing.

In 1989 the Canon Color Laser Copier (actually a scanner and a printer) coupled with various interfaces to electronic page files held out the promise of on-demand *color* printing. Although it lacked the speed and duplexing capability, it did create a de facto demand color market.

In 1991 Heidelberg introduced the GTO-DI, the first printing system to introduce the concept of on-demand color reproduction. In concept it created printing plates on a traditional offset printing press, effectively solving the registration problem associated with the four-color printing process. The plates were created with a spark discharge which then allowed them to print without water, effectively solving the ink and water balance problems associated with offset lithography. Within two years laser technology was introduced to bring the system up to commercial quality levels. In 1995 the 2-up Quickmaster-DI and in 1996 the Omni-Adast 4-up systems were introduced.

These systems whetted the appetite of the market for instantaneous reproduction of black and white and eventually for color pages. In 1993 the technology for short run on demand color printing was introduced.

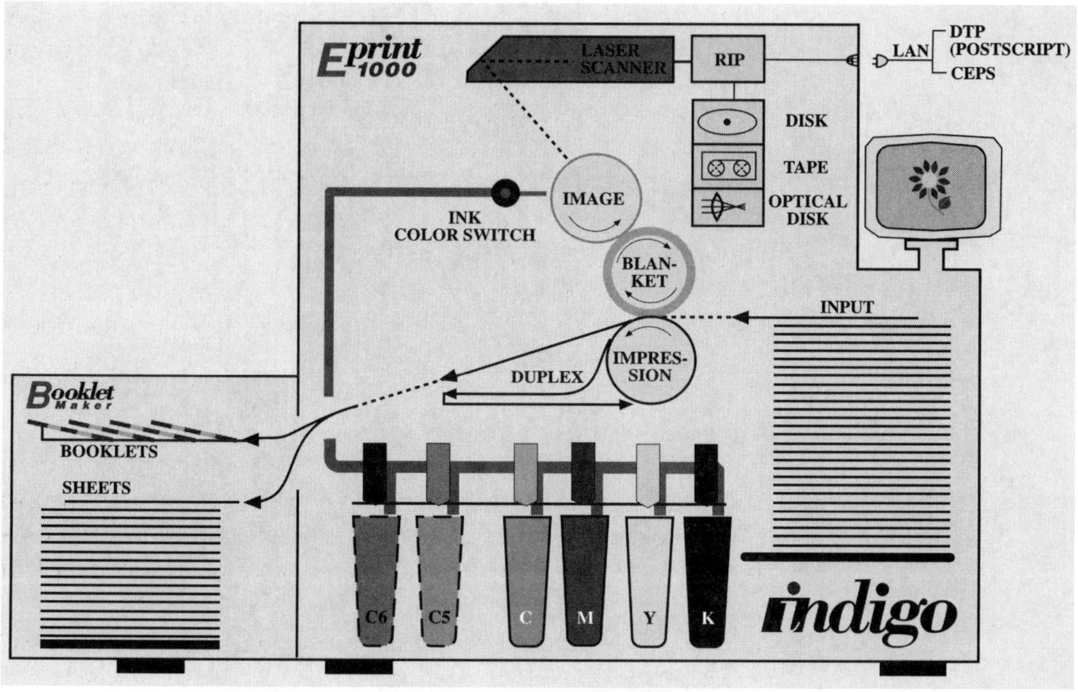

Indigo of Israel (E-Print 1000 schematic shown above) and Agfa/AM/IBM/Xeikon of Belgium (shown on the next page) have introduced integrated solutions to digital color printing. They are the vanguard of a number of systems that will address the need to produce color pages on demand.

These two systems have been first in digital color printing. The Indigo uses a liquid toner and sheetfed approach, while Xeikon uses a particle toner and webfed approach.

Both systems take digital page and document data from desktop computers or dedicated color systems. They produce sheets and documents in runs as short as one. Most importantly, they can vary the text and images on each and every page to produce a level of personalization that have not been possible before.

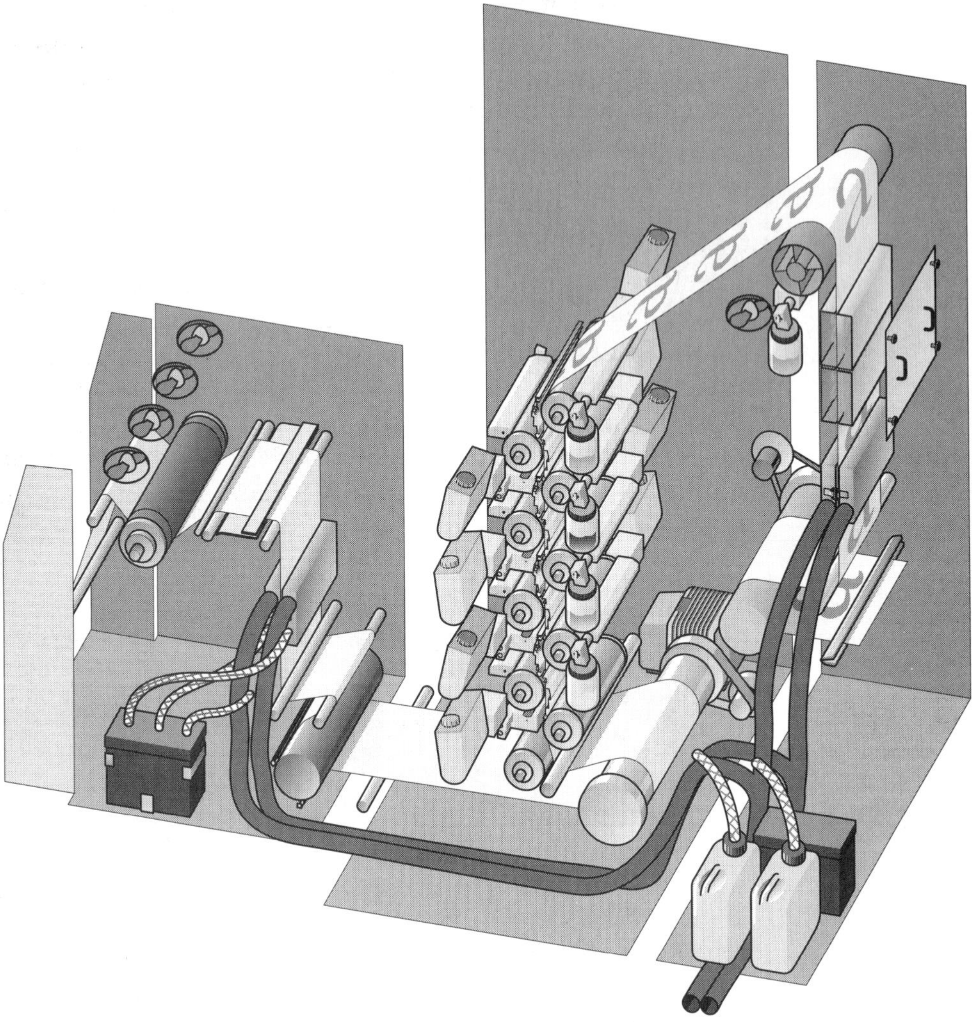

Why on demand digital printing?

The cost of paper-based publishing is rising. Print must find new production alternatives, because:

• Paper costs are rising, and may never to return to their historical levels. We must print what we need, not more.

• Distribution costs and energy costs and can never get cheaper. We must target more.

• Out-of-date information places a burden on business and industry. We must print faster.

The expectations of the information consumer are also rising. As literate consumers get their hands on a mouse and use a computer, a revolution of rising expectations will sweep publishing markets:

- Faster print turnaround and higher quality to compete with electronic media.
- Searchability of interactive products to go beyond the passivity of paper.
- Pay-per-print for information that is needed rather than information that may never be used.

Conventional printing (and even broadcast television) typify the batch manufacturing approach of the Industrial Revolution: a single standardized product produced in vast quantities with economies of scale pushed into the market. Advances in computing, communications and imaging technologies improve the targeting of publications. On the creative side, electronic design and production tools make more customized, more timely and more economical pages. On the distribution side, digital media formats make the delivery of the publication better, faster and cheaper. Publishers are crafting a new generation of publications that adapt themselves uniquely to the needs of the "Audience of One"—you.

Digital printing

Digital printing is a breakthrough technology that is changing the way the world prints and communicates. Digital presses tap the power of digital data to offer new levels of quality, speed, and flexibility. With digital printing, electronic files go directly from the design workstation to the digital press, eliminating interim steps that can compromise quality. Jobs are printed more quickly, since the digital color printing workflow is significantly streamlined and includes many automated steps.

Digital printing enables new types of communication that were prohibitively expensive—or impossible—via traditional offset. You can output short-run jobs from one copy to several thousand. You can create customized documents by varying significant amounts of printed information on each page during the press run. You can archive digital printing files for future reprinting, updating, or repurposing for CD-

ROM or online publishing. The new concept of "distribute and print" is now a reality, since you can distribute files electronically to multiple locations for local output.

The world needs information faster than ever. Businesses are changing rapidly. Information of all kinds increasingly assumes a digital format. Products are upgraded or improved quickly. Marketing continues to focus on more carefully defined target groups. Print communication continues to be the world's most widely used method, and the cost advantages of just-in-time (JIT) manufacturing are being extended to include promotion and printing. All of these forces combine to create a strong need for digital printing.

Color trade houses, service bureaus, photo labs, and other prepress and printing organizations have adopted digital printing as part of their operations. Digital printing differs significantly from traditional printing in terms of technology, workflow, and capabilities. While traditional offset printing transfers ink to paper via plates that retain ink in image areas, purely digital output is based on the electrophotographic process, which transfers microtoners to paper via carefully controlled electrostatic charges. You radically compress the printing process. Files go directly from the creator to the press, eliminating a significant number of steps from the traditional method—including color separations, stripping, plate-making, and press makeready. Digital printing allows short-run (even microrun) color printing and highly customized press runs—capabilities unavailable to traditional offset customers. And it lets customers print materials on-demand, as needed, with or without updating.

Components in a complete, integrated digital system:

Hardware: A laser electrostatically charges an organic photo conductor. Toner is attracted and transferred to paper.

Software: Designed to automate, streamline, and monitor the entire printing process—from creating job files to managing the printing queue. Personalization document customization software makes it easy to print varying information on each impression of a single print run.

Consumables: Coated and uncoated papers on webfed rolls or in sheets, plus color toners, which bond permanently to the paper.

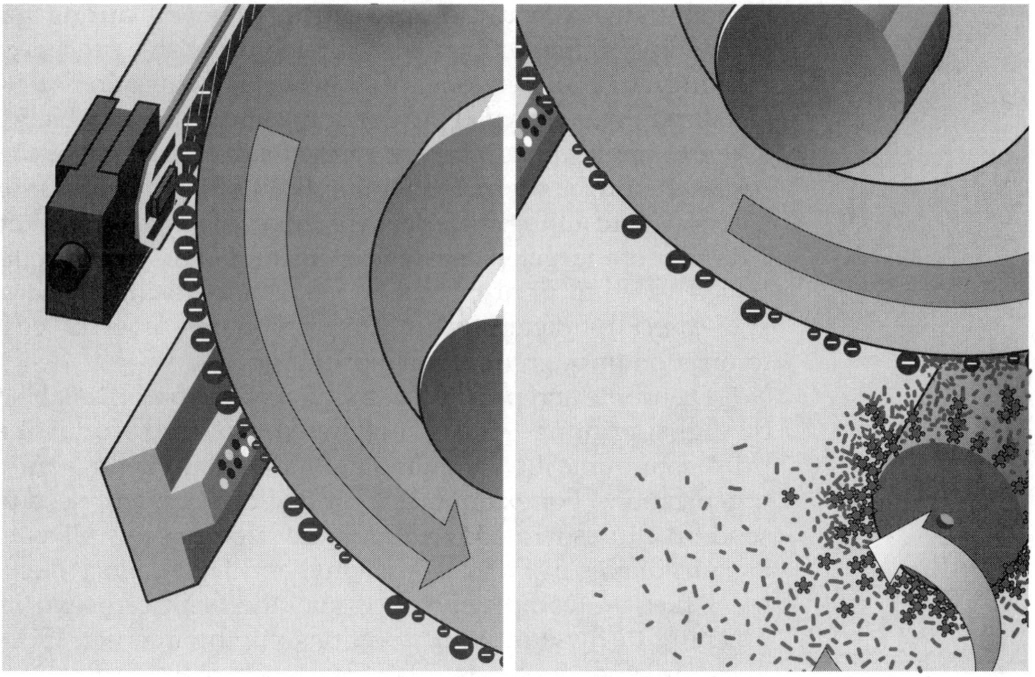

In particle-based electrophotography a corotron charges the photoconductive drum. A laser changes the charge and oppositely-charged particles of toner are attracted to the image areas.

Customized newsletters and magazines, short-run marketing pieces and catalogs, on-demand computer manuals—all of these materials and more are being produced quickly and in color with digital printing. Any application that requires high-quality, short-run color is appropriate. Here's a partial list: software documentation, hardware manuals, instant trade show publications, personalized presentation handouts, real estate brochures, insurance company mailers, textile design catalogs, retail catalogs, training binders, ad agency comps, supermarket circulars, high-end marketing brochures, test marketing materials, in-house communications, newsletters, tabletop displays, CD booklets, business cards, and large-format posters.

One core benefit of direct digital printing is its flexibility, since content can be changed, rearranged, or updated until the operator sends the digital file to the press. With personalization software you can vary printing content during the press run—eliminating the difficult and expensive process of creating several versions of the same publication. Personalized press runs of a publication—once only possible for low-end applications, such as direct mail, and in single colors—can now be produced with outstanding quality, in four colors. The resulting materials are more effective, since they have a more targeted message. And they have a high perceived value to print buyers—such as advertising agencies and direct marketers—that cannot get this level of customization through traditional printing.

Distribute and print is a communication concept enabled by digital printing—one that allows streamlined worldwide print communication required by today's marketing-based organizations. For example, a New York bank might need to send brochures to a convention in Switzerland as well as to its subsidiaries in France, Germany, and Japan. Since document files are completely electronic, the bank can send its brochure to these multiple locations via the Internet, ISDN lines, or other methods. The files reach their destinations quickly and are ready for fast output locally. This communication model reflects marked improvements over traditional methods—which required centralized printing, then international shipping, then local distribution.

Color printing technology will be as close as your desktop. Eventually the scanner, copier, fax machine, black and white printer and the color printer will merge into one productive device that will allow printout when and where you want it. Larger systems will serve larger needs and printing presses will still be available for even larger requirements.

3.

New Media

Electronic information services grew 17% in 1996 to an estimated $20 billion, slightly faster than the previous year's 15% increase. The industry continues to attract new suppliers and customers. Most major media companies are now involved in digital media publishing, whether CD-ROM or Web.

Most of the industry's revenues come from business users of information for financial management, research, marketing, purchasing, and general business administration. Online services have over 3 million business subscribers and over 4 million consumer subscribers. Consumer and small business use of online services for news, information, transactions, and bulletin boards account for an increasing share of the growth. Users of scientific, technical and professional information are also important customers.

Another measure of growth is the amount of information available in electronic format. According to Gale Research, publisher of the Directory of Databases, 221 companies produced 300 databases worldwide in 1979; in 1993, a total of 2,221 companies produced 5,210 databases. Online services increased just as explosively, from 59 to 824. In addition to the online format, 3,200 databases are available on CD-ROM, floppy diskette, and magnetic tape.

New competition is emerging as companies in other sectors of the information industry, such as Microsoft, Apple Computer, and AT&T, begin to move into electronic information services.

About 65% of information is delivered online via computers equipped with modems, but the percentage of information delivered on CD-ROM continues to grow, reaching 12% in 1993 and 16% in 1994. The commercialization of the

Internet, a global computer networking system originally used by the education and research communities, has enhanced awareness of electronic information services and encouraged the development of new services. The Internet consists of about 10,000 information networks, to which nearly 3 million computers are linked.

Progress has been made in interactive multimedia, a technology that combines computer, telephone, and television functions. This technology could attract more at-home users to information services. Personal communications networks, introduced in 1993, will be an additional means of delivering information. Every indicator shows that the electronic distribution of information is a growing business opportunity. The distinction between information and entertainment is not made, but these often overlap anyway.

Defining media

Defining "new media" as online services, the Web, CD-ROMs and interactive TV, to what extent will these alternatives alter advertising agency allocation of media advertising? Nine CEOs said new media would have "some effect."

TV is a potential victim—the average American spends four plus hours a day watching it. How much leisure time will the average American spend with TV *and* new media?

A large proportion of America's computer elite will spend an hour or two each evening in digital bliss. When they're measured and averaged along with the huge number of broadcast participants, their number could be insignificant. There are just two mass "image" media. Only TV and print give national advertisers the tools they need to quickly extend a pervasive image-building message—in color—to tens of millions of average Americans. New media may barely make a dent in the massive persuasive power of magazines and TV. In particular, TV, the non-interactive *tube*, may not yield much of its four-hour monopoly.

Remember the *Passive Emotional Reward*? TV spoonfeeds you the sitcom's laughs, the sporting event's thrills, the drama's nail-biting tension, all while you sit back quietly. But interactive media are work—you have to think and do. And emotional reward? Cyberspace doesn't have a laugh track.

CD-ROMs

Market researcher Dataquest said that 17 million CD-ROM drives were shipped in 1994, up 153% from 6.74 million in 1993. This growth in the installed base of drives has triggered dramatic growth in the sales of CD-ROM titles. Dataquest estimates worldwide factory shipments of consumer CD-ROM titles—including reference, entertainment and educational works that typically sell in the $25 to $100 price range—will rise to $450 million this year and to $2.5 billion by the end of 1997, a nearly 700% increase over last year's $325 million.

Despite such projections, nagging questions continue to overshadow the market. In the long term, there is the possibility that the development of high-speed, high-capacity transmission capabilities via cable or some other technology will undercut the advantages of CD-ROM delivery and reduce it to a transitional technology.

Even before the CD-ROM industry can face such challenges, it still must prove itself today. A disturbingly high level of customer complaints have been surfacing about high prices, disappointingly poor content quality, slow interactive speeds and monumental installation difficulties.

Another trend links CD-ROMs to online services. This approach makes it possible to create products that are rich in graphics and video, but that can be updated and changed easily—something CD-ROMs can't be. Such links to online services could be just stop-gap measures. Online services are testing cable links that allow graphics and video to be transmitted quickly and easily—diminishing CD-ROM's appeal.

Private systems

Companies and organizations have private networks and do not open them to the public. Along with the traditional bulletin board functions, private systems sometimes allow users to communicate privately, perform specialized computing operations or access specific data.

Computer Bulletin Board Systems (BBSs)

BBSs were the precursors of commercial online services and allow users to set up accounts on remote systems and access basic system features. The distinction between BBSs and online services is not always clear. Most BBSs are operated on small personal computers with a single phone line and have a relatively small number of users. There are usually no access charges since the system operator, or sysop, is running the board as a contribution. BBSs emphasize electronic mail, message bases, and software trading. Systems also host games, and some offer user chat. Because of the large number of bulletin boards in existence and the ease of setting them up, bulletin boards tend to cater to a wide range of specialized interests. City, state, and federal government agencies run bulletin boards as public services. Sysops have wide discretion in shaping their boards, in limiting membership, and in establishing rules for participation.

Commercial online services

Remote systems most familiar to Americans are the commercial online services. These are large computers that often have the capacity to host thousands of users simultaneously. Some of the more popular services include Prodigy, CompuServe, GEnie, America Online, and Delphi. With a computer and a modem plus the proper software you can have a personalized account. Each time users log on to the service, they enter a unique name and password combination that identifies them to the host system. Once the account is established, the user can surf the system selecting activities from a series of available menus. The number and nature of activities varies from system to system. There are comprehensive databases, electronic mail, libraries of electronic publications, entertainment activities and more. There are vast collections of software, online conversations, messages posted by users, and current news and financial information. You can make airline and hotel reservations or shop in online stores. A sense of community permeates system subscribers, who may spend hours interacting with the service and each other.

Computer networks

Computer networks are collections of interconnected computer systems that have the ability to exchange information among one another. Systems from each of the previously discussed categories might be part of a network. The most important computer network in the world is the Internet, which has been called a *network of networks.*

The new marketplace

The largest category of new media offerings is retailing. The lack of social interaction of the online home-shopping experience has led to these programs not being as engaging as the consumer might have expected. Most of the multimedia activity supporting shopping is now found in retail stores, enhancing the experience of more traditional shopping by providing point-of-sale, highly polished, interactive marketing information. In fact, multimedia kiosks support or replace salespeople in many cases. A kiosk can be *re-trained* easily and can assist in providing quicker and more reliable transaction service. Also, virtual reality systems with full 3D visual and touch effects are used to provide consumers the ability to *fully experience* a big ticket product before the purchase. These virtual reality systems also contribute to attracting consumers to stores due to the entertainment value of these experiences.

The Net allows businesses real-time feedback from their customers. Thus, customers are able to exercise their demand for control and become actively involved in the design and manufacture of their purchases. Businesses, in an effort to meet these demands, have upgraded their manufacturing systems to allow for this requirement.

Shopping will change as merchants remake their stores into mini theme parks. Consumers experience big ticket products in traditional stores, but often complete the transaction via the Net using cyber dollars. In addition, a large number of many households' financial transactions as well as purchases of basic goods and services (e.g. food, household supplies, and entertainment) will be handled over the Net. This has contained the *threat* of retail *slotting fee blackmail,* and has provided unlimited shelf space. Also, it has provided market research companies access (with the consumers' consent) to the buying habits of large cross sections of consumers.

This valuable information has been made available for a small fee to the many small companies that market on the Net and is responsible for these companies' ability to anticipate the consumer's desires. As the logic of *information content* delivery changed, and with the hopes of attracting mass audiences lowered, marketers realized that their traditional electronic advertising method was changing. Now consumers select—and pay for—their interaction with the Information Node at home. This is not an environment in which the *uninvited persuader* works. Marketers, no longer able to use mass mailings due to rising costs and environmental pressures, and screened out by Information Nodes, have turned to providing content as well as product information on the Net. The trend began with *infomercial programming*, and has now become far more pervasive on the Net.

The marketplace of the mind

What is interactive advertising anyway? Is it TV? Is it kiosks? Is it CD-ROM? Is it the Web? There is a word for it. It's called "selling." Interactive advertising is the perfect selling medium—a marketplace of the mind.

It's not TV. TV is too passive. You can't get people involved in a sales process by shouting something at them over and over again. TV commercials rely on the message and repetition . . . and a passive viewer.

Information is king (or queen). If someone is really interested in a product, they want information about it, which is the point at which you can sell them. Interactive is collateral advertising, like direct mail.

Good direct mail draws you in, gets you really involved, makes you an offer you can't refuse, persuades you to get out your credit card. There are even attempts at interactivity, as in getting you to paste a stamp to a card or check a box on a business reply card.

Copy is key. You're going into an informational medium, words must get your message across. A picture may be worth a thousand words, but which thousand? Use graphics and copy the way you would in a good print ad. Focus the message, sell in the copy, make your point.

So what if you lose control. The automobile provided individual empowerment: Go where you want, when you want. The same is true of interactive advertising. Don't try to control the flow. Let users look at what they want, go where they want, interact at their own pace. Make sure your program is wonderful wherever they go.

New media

New media refers to all new publishing formats that apply computer processing, communications and imaging to meet the needs of information consumers. Synonyms abound: computer publishing, new media, interactive media, multimedia, cybermedia, hypermedia, computer-mediated communications and electronic publishing. Digital media can deliver information as:

- *Substrate-based:* traditional print as well as "customized" print: selectively bound magazines and catalogs, ink jet-customized pages, demand printed technical documentation and books, distribute and print approaches, and both broadcast and interactive fax publications.
- *Media-based:* alphanumeric and image databases on magnetic media or optical media, and CD-ROM, Digital Video Disk, CD-I and PCMCIA cards.
- *Communications-based:* commercial online services, Internet and World Wide Web publications, network and groupware products, interactive television and wireless communications systems.

The fundamentals of publishing don't change just because the final product is print or digital. Digital media is just like conventional media and the same questions must be asked:

- Marketing: Does anyone want this product in any form? Can it be sold or distributed through existing channels? Are there enough people who will pay the right price?
- Editorial: Will the publication communicate better using the digital media tools? Are there advantages of customization, timeliness, comprehensiveness, searchability, economy or transaction?
- Promotion: Will advertisers support electronic advertising; will they profit from direct response?
- Design: Will the design elements of digital media enhance the product or information presentation?
- Production: Is the data in electronic form? Can it be con-

verted from the conventional product?

Too often digital media projects start because of all the wrong reasons: the file is on disk (you assume), the marketing department thinks it wants it, and the art director just took a one-day course in multimedia design. Interactive is not enough. Sound and animation are not enough. Will it improve the process of communicating *your* information? Create a totally digital product and you gain:

• Customization: Hyperlinks, reader profile filters, information agents and database marketing approaches fine-tune a publication to the "audience of one."

• Timeliness: Digital media telescopes schedules for both editorial and advertising. It's really fast.

• Comprehensiveness: No print-based library can compare to the mountains of information available on a CD-ROM or on the mother of all publications, the Internet.

• Searchability: Search through a database and retrieve your data faster than you can flip through a print index.

• Economy: Audiences expect digital media to be more economical to purchase in pay-per-view configurations by the issue, by the article or by the paragraph.

• Transaction: You can have the reader talk back through a newsgroup, order directly from an online catalog or buy products via a CD-ROM.

New media is just like old media. Only different.

4.

INTERACTIVE MULTIMEDIA

Multimedia is the synthesis of digital content and form for information presentation, exploration, and dissemination as applied to education, business, and entertainment. Many words are currently being used to describe the meaning of this definition. Cybermedia, Intermedia, New Media, Hypermedia, Interactive Media Design, Interactive Multimedia, Electronic Media, Computer Graphic Design, Unimedia, Multimedia and Digital Media. They all essentially arrive at the same place: the distribution of products in electronic form, delivered on a medium (disk) or over a telephone line or network.

Technology is forcing many disciplines into new forms, shapes and relationships. A prime example is the present convergence of the entertainment, communications and computer industries. And there may never have been a time when change was more rapid and dramatic. Government, industry and education are primed for participation on the "information superhighway." *Multimedia* is the computer integration of different forms of communication media for the presentation and dissemination of information and knowledge, for education, entertainment, simulation, and promotion.

Presentation: Multimedia can be used to present information in new and more interesting ways. It integrates color, graphics, video clips, animation, type and sound. It provides new tools for conveying ideas and making points.

Simulation: Multimedia can be used to simulate control panels, gauges, and visual displays so that users can be trained in certain operational procedures before approaching an actual system.

Promotion: Multimedia can bring products "alive" and

allow potential buyers to learn more about them through sound and video. More importantly, the potential buyer can interact with the program to get the information they want.

Education: Multimedia can be designed to be interactive and thus work with the student in the process of discovery. The program can help them navigate through the information in some orderly fashion while appealing to both sight and sound.

Entertainment: Multimedia is "MTV publishing"— and for some in that it can be flashy, superficial, and confusing. But, it can also be games and some games can actually be educational.

The confluence of communication systems

Multimedia is a technology that provides tools for information delivery, but there must also be a reliance on content. Multimedia brings together a number of approaches to communication:

Technical: The confluence of these six areas of technology that influence human communication, organized to present information: Audio, Color, Computer, Photo (Image), Text (Type), and Video.

Instructional: Multimedia will play an important role in education and must integrate the following philosophical and theoretical areas: Communication Theory, Instructional Technology, and Learning Theory.

Developmental: Multimedia requires that technical and philosophical considerations meet: Content Development, Information Technology, Presentation Development, and Program Development.

Creative: Multimedia is a creative process: Animation, Graphic Design, and Interactive Design.

Some societal and technological trends are evident:

- The merger of telephone, television and computer-based technology.
- The advent of hundreds of channels of cable television.
- The growth of distance learning.
- The application of computers in teaching and learning.
- The development of electronic publishing items, like catalogs, directories, and reference materials.

- The dissemination of information in print and non-print.
- The development of data highways, which will link academic, commercial and governmental users.

Each of these areas increases the access of human beings to increasingly larger bases of information and knowledge. To utilize this information and knowledge more effectively, it will have to be organized and presented in new forms which take advantage of the increased capabilities of the technology. These capabilities will have to be applied by a new breed of information developer, who will apply conceptual, aesthetic, technical and integrative competencies to organize and design information for presentation.

The new information developer

This new information developer will shape content in exciting and innovative ways. With a background firmly rooted in typography, aesthetics, design, imaging, and presentation skills, and using the computer as an integrative tool to assemble pages and screens, they will apply new approaches to information packaging.

Thus, there will be two levels of application:

1. Multimedia producers: they will convert content from others into new form. They will work in-house or as commercial services.
2. Content developers: they will be publishers or teachers of others with content and they will either master the tools to put their content into form or they will utilize the services of developers.

There will be a need for a large pool of people trained in this technological approach. They will have to learn these skills somewhere.

New technologies—ranging from animation to virtual reality to real-time audio—are converging to transform the World Wide Web from a two-dimensional medium into one that encompasses multimedia and true interactivity. Even the jargon is changing. Companies that today are rushing to create home pages will soon be able to create *spaces* that defy the physical boundaries that a print term like *pages* implies. The Internet is becoming a multimedia publishing vehicle.

Sun's new programming language called Java and companion browser called HotJava bring instant interactivity, audio, video and animation to Web sites. Sun's tools follow on the heels of Silicon Graphics' WebSpace, a browser capable of displaying three-dimensional images. It works with a new Web programming tool called Virtual Reality Modeling Language that enables users to "walk" around a virtual shopping mall, enter a store and view a product from all angles.

The creative challenge of interactive media

The new frontier of interactive creative—and its potential impact on the way agencies and marketers approach consumers—is a hot topic. You can't just reproduce TV ads for interactive TV. Those who believe cyberspace will make dinosaurs of magazines should check out their local newsstands. Magazines about the Internet and other new-media technologies are hot.

New media is a collaborative process. This chart show that projects are usually divided into various parts:

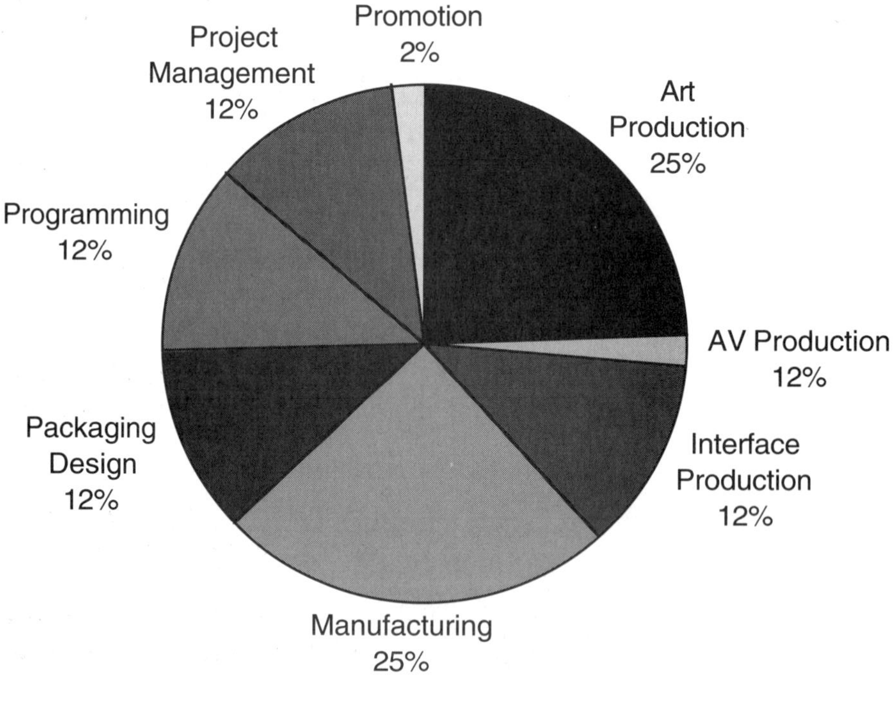

Romano's 21 rules for interactive multimedia

1. Capitalize on the users' frame of reference. Use icons and symbols that they understand. What can you do with an arrow that is new and creative? How do you indicate "Next Page" without saying it? *Icon: A few thousand pixels doing the work of one or two words.*

2. Be consistent with the outside world. Think of how normal people would normally act. Can you go beyond up, down, back, forward? Space travel? Time travel? Get me to the next page on time.

3. Have fewer richer screens. Don't waste my time with a line of type, a glimpse of art and thou beside me in the wilderness yakking.

4. Provide a common look and feel. Not the uniformity of the military but a family resemblance to the milkman.

5. Provide navigational assistance. Provide a traceable path to and from wherever. You are here! Where's that?

6. Be consistent, consistently.

7. Efficacy: How will the user react? . . . interact? Overact? What is clickable? What is not.

8. Pretend that the program is a game. Make it fun, mysterious, intriguing.

9. Aim for conceptual integrity. In other words, give the user some element of commonality on each screen so as to tie the screens together.

10. Give the user control of a complex program in a simple manner. Put the user in the driver's seat.

11. Reduce the users' need for rote memory, training and documentation. We hold these icons to be self evident.

12. The purpose and use of a control should be obvious from its form. No user should have to try to figure out what it means or how it works. Thoughtless interactivity?

13. Good programs offer clues to location and navigation. Leave electronic crumbs for the user to follow. Don't be too cute. Don't be too sophisticated.

14. Bad programs confuse the user. Bad users confuse the program.

15. Good programs get the user involved. Keep your sense of humor . . . somewhere between Robert Benchley and the Three Stooges.

16. Make the program and its operation predictable. If A = B don't let it = C later on.

17. Provide rich and subtle feedback. When the user performs an action, let them know you care enough to send the very best. Users love a response to an action—it is an acknowledgment, a reward even.

18. Imagine a perfect solution and temper it with reality. Create the realistic solution and temper it with imagination. Have one foot in the real world and one foot in cyberspace.

19. Create escape routes so that users can back up or quit or just give up easily. This way out.

20. There is a thin line between real creativity and the need for psychiatric help. The real pioneers in multimedia design will not be recognized until they are dead.

21. Readability isn't everything; it's the only thing. Text is not art, maybe.

5.

Portable Documents

A commercial printer "prints" automotive parts catalogs to CD-ROM. An industrial catalog firm distributes a 2,000 page catalog on disk—via their printing company. A computer manufacturer provides all technical documentation on CD-ROM. The traditional print version was converted and produced by their printer. A large printer spends more on blank disks than on paper.

Printers do more than print on paper. They convert and distribute information. Once you see that clearly you are ready to join the majority of innovative printing firms in the United States who are deriving revenue from digital products and services. And, by the way, they still provide print-on-paper products and services.

Most of what has happened comes about because of the confluence of two technologies: the very high density read only disk and the portable document formats that let you create documents that can be read but not changed. Put them together and you have a digital publication—one that requires a computer to "read" it.

Documents are not all produced with the same program. To read your memo created in Microsoft Word, I would also need the same program (and probably the same revision level and the same fonts). This is all right if everyone uses Word, but that is not the case. Documents are created and/or produced with word processing, spreadsheet, database, graphics/illustration and desktop publishing programs.

Certainly one could release the text as simple and standard ASCII code, which just about any program could import. But ASCII would not give me the type, graphics and page formatting. Users demanded that the content and the form be preserved and distributed, with some common approach to viewing and accessing the document even if you did not have the originating program or the fonts or even the same computer platform.

The solution was a portable document format and a standardized viewer.

The good, the bad and the digital disaster

The underlying concept of document portability is that of printing to a file. As an analogy, take a sheet of paper with text and graphics on it and fax it. The sending fax converts the page images to dots and the receiving fax prints them out. If you have a fax capability from your computer, a program takes the page image, converts it to dots and sends it to the receiving fax for printout. Now, save the last file we created—a representation of the page as dots—and instead of printing it to paper, put it on the screen.

But there is something missing. Like any fax image, there is no underlying *intelligence* for the text. You could not search through it because it does not know an *a* from a hole in the paper. Searching is something you want.

Most portable document programs save a bit map of the page as it appears on the screen, the underlying ASCII text, and the font data. That makes portable documents like some of the portable TVs of the 1950s when you had a unit that weighed hundreds of pounds with a handle on the top. Even with compression, most portable documents are six to ten times bigger than the original file. By having the text in ASCII you can search for words and phrases. This is a major advantage over print. The material in a book or catalog is not information until you find what you want. Search and retrieval lets you find what you want. Another advantage of ASCII is that you can cut and paste text. You do not get the formatting but you do get the words. The bit map capture of the screen provides the print image. Usually, this is saved at 300dpi and that is the best you can do for printout, no matter what resolution the printer is at.

Typographic fax

Adobe's Acrobat is somewhat different from above. It is based on PostScript, the page description language that has become the mainstay of printing and publishing. Acrobat saves a PostScript representation of the page. It can then output to any PostScript printer at whatever resolution the printer has. However, for the sake of efficiency one version uses a font substitution approach based on Adobe Type Manager and Multiple Master fonts technology. You will get type that has the same width and height but not the same style as your original document.

Another version will preserve the font data but it increases the file size significantly. But, this latter approach provides the most important aspect of document portability—the ability to print out at any resolution.

This means that the receiver of the digital page could someday have a high resolution color printer and print out pages as needed in a remote location. Pages could be created in one part of the world and then sent to a printer who uses the data to make high resolution films for printing. Ads could be created at ad agencies and sent to publications.

This idea is not a new one. The advent of low cost, better quality printers could mean that instead of printing one million pages on a press, one million presses print one page. End users will only print out what they want, whether it is the executive summary of a thick report, or the page of a technical problem that answers your question, or the legal citation that supports your position. All examples are in current use.

Compression and viewing

Given the large file sizes that result from any of these approaches, compression is an absolute necessity. Some programs automatically compress when the file is converted and then decompress when it is viewed.

The viewer is an important issue. There are usually two parts to the program: the converter and the viewer. The converter, or whatever, it may be called, converts pages from virtually any program that runs in the desktop world. The viewer is the program that lets you open the file on another machine, navigate through the document, search and retrieve, etc. There are usually two approaches to viewers. Some programs insert a mini viewer with the document file so it carries along its own ability to open, and it is free. These mini viewers do not have a full set of tools and thus a more expansive version is available—at a price—to provide them. In all cases, there is only one kind of viewer.

It is safe to say that the viewers that are most commonly and easily available will make their programs popular and prolific. At present, every supplier of these programs has or shortly will have a converter and viewer that runs on Macintosh and also Windows. Some will also run on Unix.

The reason that portable documents are so important is the lack of standardization in the creation and production of pages. Standardization was imposed by establishing portable formats. Now there are four major companies with competing products. They are Adobe Acrobat, Farallon Replica, No Hands Common Ground and WordPerfect Envoy. For the printing industry, Acrobat holds the most promise because of its ability to print to film, plate, or digital press.

How printers work with portable documents

Portable documents begin with pages that were created or produced at the desktop; that is, with programs that run on the Macintosh or PC under Windows. Printers routinely receive QuarkXPress documents, for example. They usually output the pages to film and print. They can now offer their clients additional services by converting the QuarkXPress files to a portable format and mastering and duplicating CD-ROMs. The client can have any distribution approach they want and the printer can get a piece of the action.

	Search and Retrieval	Font Support	Viewer Approach	Compatibility Cross Platforms	File Creation
Adobe Acrobat	Excellent	Excellent	Good	Excellent	Very Good
Farallon Replica	Good	Poor	Good	Good	Good
Microsoft	Good	Good	Poor	Poor	Good
No Hands Common Ground	Excellent	Good	Excellent	Excellent	Excellent
WordPerfect Envoy	Good	Good	Good	Good	Very Good

They call it *cross-platform compatibility* but it really is **in**compatibility with different *platforms,* or computer systems (Mac, DOS, DOS/Windows, Windows 95, Warp OS/2, and UNIX) trying to talk to one another.

A user of nationwide newspaper advertising and the Associated Press have developed something called AP Adsend. You create a newspaper ad on a Macintosh or PC and save it as a PostScript file, distill it into an Adobe Acrobat portable document file, fill out an onscreen delivery ticket specifying which newspapers are to run it and when, then transmit it in compressed form to the AP. They will transmit it directly to the designated paper or papers.

The ad is received by the newspaper's computer looking just the way you created it. They position it into their desktop file for the issue. No film. No overnight delivery service. Adobe Acrobat does it all. Introduced in 1992 as a pioneer product in cross-platform document distribution it uses Adobe PostScript to create a common denominator usable by most major computer systems.

PostScript is device independent and Acrobat is platform independent—the document is readable and usable by any system that has a Reader program. In AP's, case Acrobat sends the images (color and b&w) and the type and the page layout down—as a portable document. The fonts and the document travel in a bundle called a portable document format, or PDF.

This technology goes beyond the transmission of newspaper ads—it allows access to entire electronic archives of information. The browsing, searching and reading possibilities are extensive. It is a giant step for document independence, where software creates a giant electronic manila envelope filled with just about anything from pages to sound to video. With cross-platform interoperability and sufficient bandwidth for video/audio/text transmission, an advertiser will be able to send ads and commercials directly to publications, broadcast media, or even directly to online service subscribers.

Paperless office, huh?

Portable document software came into being in 1993 with the introduction of No Hands' Common Ground and Adobe Systems' Acrobat. Developers claimed they were the answer to paper documents needlessly killing trees, piling up unused in warehouses or filing cabinets, and causing frustration among business people unable to find the document they needed in the sea of papers around them.

You can easily move documents around by e-mail, server, or disk, and you can't assume that everybody has the right fonts on their systems, or that they have the right program to open your document, or even (in a cross-platform environment) the right setup to receive the document. You could spend a lot of time and money installing the same software and fonts, plus the requisite extra hard disk space and RAM, on every system to allow document portability—and then train people on each program used to create the documents in the first place. The whole idea just could not work and technology created the problem, so technology could solve it.

This problem was obvious, and soon several companies offered a software solution: portable document programs. First came No Hands Software's Common Ground. Adobe Systems later shipped Acrobat which legitimized the market. Farallon Computing followed with Replica.

By creating electronic documents that carry all the needed components—fonts, graphics, and a viewer or reader to view and print the document—this type of software could eliminate the cost and time of printing, distributing, and storing paper copies, while adding the ability to find text and link multiple documents so that information could be both more accessible and more dynamic. And very portable.

Although known in industry parlance as portable document software, what distinguishes the technology is not its portability—e-mail and networks make any document portable, after all—but its autonomy. Portable documents are self-reliant documents that remain intact no matter what their computer environment.

Whether for use within a company for document exchange and distribution, or for use on bulletin board systems, CD-ROMs, and fax-back services for user-requested documents, the possibilities for portable document software are endless.

But the first generation didn't live up to its promise. The initial programs had problems, not just rough spots, in the capabilities they offered. For example, Acrobat had no distributable reader, thus imposing an additional cost for those wanting to distribute documents widely. Replica had great difficulty with fonts and graphics, resulting in documents that bore little resemblance to the original.

However, several companies have fixed all that by offering a second generation of portable document tools: Acrobat 2.0 tries to go the next step, and No Hands Software promises to follow suit with Common Ground 2.0. In addition, two new players enter the fray: Novell's WordPerfect Envoy and Apple Computer's portable digital document (PDD) feature in System 7.5.

Portable document software is aimed at three types of users, whose needs differ:

• Collaboration—Workgroups need tools that let them work in concert on documents, so they can trade comments and incorporate elements of shared documents into their own. High fidelity, small file sizes, annotation tools, and search capabilities are important. An affordable annotation-and-search tool is required.

• Repository—Corporate communications and publication services departments want tools that let them put common documents into a repository that anyone can access—not just to view, but also to search. Capturing text that can then be highlighted, copied, and searched by recipients—but not changed—is most important.

• Distribution—Digital publishers need tools to create computer-based versions of documents for read-only use. It's the electronic equivalent of a fax, useful for the distribution of newsletters, price lists, and forms, for example. High fidelity and small files are most important.

Most PDFs create shared documents for lookup (such as corporate phone directories) and markup (such as draft memos circulated for comments). The differences between programs come down to price, user interface and complexity.

Acrobat comes in several configurations. The $195 Acrobat Exchange lets you create, annotate, and view documents. The $595 Acrobat Pro includes Exchange and Distiller. Distiller converts PostScript print files into portable documents. The $1595 Acrobat for Workgroups version has ten licenses for Exchange, one for Distiller, and one for Catalog, a Windows-only indexing tool that creates indexes to let you search across multiple documents (normally you can search only within a document). Acrobat for Workgroups includes both Mac and Windows software for Exchange and Distiller so you can use it for Mac-only and cross-platform networks.

Acrobat 2.0 includes all of the software necessary for creating electronic documents. This software includes Acrobat Exchange and PDF Writer. These are used to create electronic documents. The Acrobat search tool is also included for full text searches. The Acrobat Reader is included and is used for Macintosh, Windows, DOS, and Unix systems.

Like its competitors, but unlike Acrobat 1.0, all of the Acrobat 2.0 or later packages include the Mac and Windows versions of Reader, a program you can freely distribute so recipients can read, search, and print Acrobat documents. Acrobat requires the use of Adobe Type Manager, a font-fidelity engine, in version 3.8 or later and bundled in the Exchange, Pro, and Workgroups packages. Acrobat's Reader won't run unless you have a full version of ATM installed. It uses the font representation functions of ATM to render typefaces on screen and for printout.

Envoy initially came on one disk. It lets you do essentially what Acrobat Exchange does, and Envoy's Reader is similar to Acrobat's Reader, except that you can embed Envoy's Reader in your document (adding about 360 K), rather than distributing a 1.2 Mb file that the recipient must install, as you must with Acrobat. Envoy also has a stand-alone viewer, available in Mac (650 K) and Windows (200 K) versions, which is helpful in instances where you don't know what platform your recipient uses. The stand-alone viewer is also preferable in a cross-platform network: Windows users can't launch an Envoy document that contains an embedded Mac viewer; they must have the stand-alone Windows viewer, which will ignore the embedded Mac viewer. Macintosh users face the same problem with documents using an embedded Windows viewer. Envoy does not require the use of ATM, but you'll want ATM 3.8 or later anyhow because of Envoy's poor font fidelity.

Envoy can neither embed fonts nor emulate their appearance. Instead, the program substitutes the closest font it can find in the font list.

Acrobat is better at font fidelity, but it is hardly perfect. Acrobat occasionally converts fonts it doesn't know how to handle—typically symbols and non-Roman fonts, and sometimes fonts created by programs like Macromedia Fontographer and Ares Software FontMonger—into bit-maps, so you can't search any text using them. To get around this you can embed the problem fonts in your document (via Page Setup) before printing to Acrobat format, or you can print to a PostScript file with fonts included and use Distiller to generate the Acrobat document.

Most PDFs let you insert bookmarks, view thumbnails of your document's pages (which you can rearrange) so you can quickly decide where you want to go, and add electronic Post-It notes (with customizable colors and names, so in a multiuser setting, you can tell who wrote what).

In addition, both programs can create hypertext links between sections of your document, only they don't change the hypertext color—you need to color that text in your originating word processor or page layout program. But both programs put a color frame around the hypertext item.

The two programs set up links differently. Envoy has a palette of hypertext jump icons you can insert (including arrows); in Acrobat, you have to add any such graphics in the originating program. But Acrobat lets you control the view the user gets of the destination page when he or she double-clicks on a hypertext link, and Acrobat lets you link across documents, not just within them. Envoy is not smart enough to adjust the link destinations after repagination.

To centralize documentation—everything from price and other lists to manuals, from product specs to press releases—and make them accessible to your whole company, or outsiders, Acrobat Catalog is the answer.

The Catalog program in Acrobat for Workgroups generates search indexes that let users look for text in any collection of documents in a particular folder and its subfolders, which makes it easy to find all related documents. Acrobat uses the Verity Topic search engine, a tool that does sounds-like and grammatical-variation searches. The Verity engine also estimates how good a match each document is to your search parameters.

The Verity search tool comes with Exchange but not Reader. The Verity Topic search feature in Acrobat is a plug-in that Adobe bundles with the product. While Adobe expects many plug-ins to be developed for niche markets, one broadly useful feature is an OCR capability called Capture that lets you scan documents, do character recognition, and then create an Acrobat document based on the text.

If your goal is one-way distribution to a wide range of computer users, you can't necessarily count on Acrobat or Envoy to handle this distribution need. To use Acrobat, everyone must have ATM installed; to use Envoy, everyone should have the same fonts. Although, Acrobat will let you embed all the fonts,, which increases file size. For distribution, you want software designed to produce files that are faithful to the original, that anyone can read or print, and that recipients can optionally search and copy content from.

Common Ground has been around longer than any of the competition. Replica offers features similar to Common Ground's but Common Ground 2.0 uses a font embedding engine (Bitstream's TrueDoc).

Document control

You can prevent recipients from printing files, copying text, or copying graphics—or any combination of the three. That's important for publishers, who often want users to be able to excerpt the text but not the images (because of different copyright agreements reached with authors and artists).

Both Replica and Common Ground offer basic text-search capabilities (by word, for example) that are sufficient for distribution of individual documents.You can embed a Common Ground viewer, which lets anyone view or print a Common Ground document, in your document. Best of all, and unique to Common Ground, you can choose whether to embed a Mac or Windows viewer, so you can do your authoring on the Mac and distribute fully self-contained portable documents to people on both platforms. And since it's the smallest embedded-viewer size around—just 90K for the Mac and 100K for Windows—you can afford to routinely embed the Common Ground viewer. Of course, you can also distribute the Mac and Windows viewers separately, and not embed them in your Common Ground documents. This last scenario is best for BBS operators or online services who don't want to boost the file size for each document unnecessarily.

If you want your recipients to search for text in your files or copy out text or graphics, they'll need the full version of Common Ground installed. The embedded and stand-alone viewers just let them print and display documents. Replica also embeds a viewer, which amounts to sending along the full Replica program—at a cost of about 400 K. Both programs let users select text in columns, although Replica's method is somewhat simpler.

Additionally, to ease production, both programs let you drag and drop files onto creator programs, thereby launching the files' creator and generating the portable document, using default settings. It is easier to create the portable documents from within Common Ground or Replica (both programs offer keyboard shortcuts to skip going to the Chooser to change the target printer), since you can then control output settings from within the host program.

Common Ground captures the text that can be highlighted, copied, and searched when you print, and it saves a bit-map version of the file so fonts reproduce decently if the recipient doesn't have them (Common Ground uses the actual fonts if the recipient does have them). As with other portable document programs, having the font-emulation features of ATM augments Common Ground's native font rendering capabilities.

While Common Ground limits you to printing at 300 dpi, Replica lets you embed TrueType fonts in your document, so recipients can print text at their printer's top resolution. But Replica can't embed PostScript fonts. Both programs let you downsample your document's resolution to 72 dpi (for on-screen viewing), which saves file size. Common Ground and Acrobat also let you create files using JPEG compression to further reduce image size, but as with all JPEG compression, the greater the compression amount, the less faithful the image will be.

Portable document software is only now living up to its promise. At present the clear winner is Acrobat.

Adobe Acrobat gives instant access to documents in their original form, independent of the type of computer used. With Adobe Acrobat the user can view, navigate, print, and present any portable document format (PDF) file.

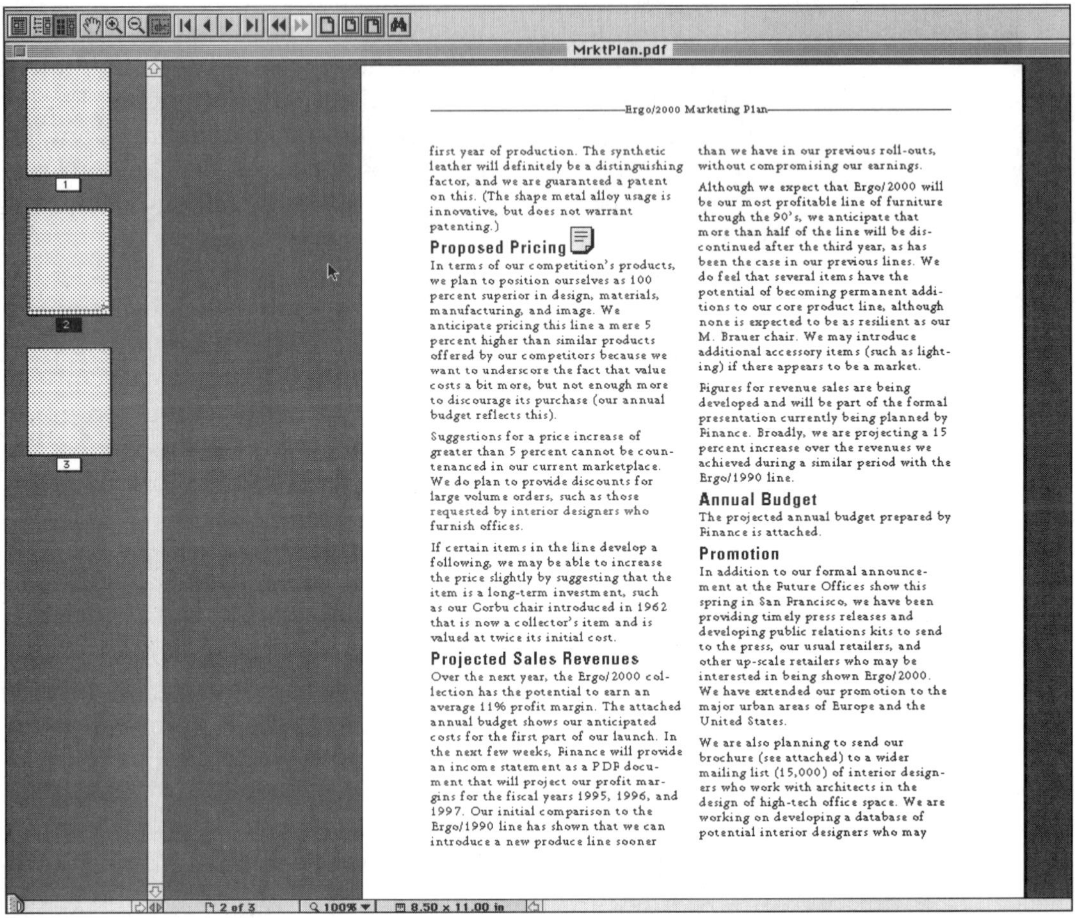

Digital ads

The Committee for Graphic Arts Technologies Standards (CGATS) has announced that Subcommittee 6 of CGATS has selected Adobe System's Portable Document Format (PDF) as a basis for a standard to address the industry requirement for digital distribution of printing data.

CGATS formed SC6 in response to a need to develop standards to permit transmission of digital ad data electronically. The Digital Distribution of Advertising for Publications (DDAP) Association has been lobbying for the use of industry standards to allow digital ad distribution for some time. SC6, in response to this need, is developing standards that will allow for digital ad distribution as well as facilitate digital data exchange of composite page data for other print applications.

SC6 is finalizing the user technical requirements to summarize the needs of the industry. Based on these requirements, two standards will be written. One will focus on the exchange of raster data based on the work of IT8 and ISO TC 130. TIFF/IT-P1 (ANSI IT8.8-1993 and ISO 12639) will be the basis for this work which will allow the exchange of rendered pages targeted to a variety of printing processes in a transport-independent format.

The challenge of transferring a combination of object-oriented data and raster data concurrently has been more difficult to resolve. The solution will be a second standard, titled "Graphic technology exchange of digital page data in final form using PDF." The standard will detail a transport independent exchange mechanism based on Adobe's PDF data exchange format. The PDF format is open-ended and allows TIFF/IT-P1 raster data objects to be packaged within a PDF file and linked to PDF objects. The application software that accepts the PDF, such as Adobe Acrobat, could view the TIFF/IT-P1 information with an appropriate TIFF/IT viewing plug-in.

The PDF was developed to meet the needs of many industries transferring, in a device-independent form, data targeted for rendering on a variety of different rendering engines from personal computer screens to printing presses. We are working closely with SC6 to define a process that fits this specialized application. PDF is a file format used to represent a document in a manner independent of the application software, hardware and operating system used to create it. PDF, the Portable Document Format, has been published as a publicly available specification by Adobe Systems.

Some of the key elements to be included in this CGATS standard:

- Fonts will be included with the PDF to assure rendering

with the exact same font and font metrics.

- Raster data will be accommodated by incorporating or calling external files, including TIFF/IT-P1.
- Data to be exchanged will include single colors or CMYK process colors with color rendering.

The initial requirements for these standards are based on The Digital Distribution of Advertising for Publications (DDAP) User Requirement Specification—Version 3.2, published by the DDAP Association.

About Adobe Acrobat

What are the salient features of Acrobat?:

1. View documents with guaranteed page fidelity. It is not necessary for the user to buy the application.

2. Time is saved by sending information over e-mail.

3. Collateral and technical documentation can be stored electronically and accessed instantly, this is helpful with sales and customer service groups.

4. Creating and distributing PDF files can be accomplished easily.

Adobe Acrobat is "all of the benefits without the paper." The software is the fastest and lowest cost way to move beyond paper and lets the user have color, graphics, fonts, easy navigation through files, and easy printing. With the efficiency of the electronic access it is possible to send a PDF file and a free Acrobat Reader to any Mac or Windows user. The Mac or Windows user is then able to view the document in its original form using the software and hardware that they already have.

It is possible to distribute and manage documents electronically without a large investment in time and money. The documents will appear on the screen at the highest possible resolution. Acrobat also can print perfectly in black and white and output document sizes up to 17x22 inches.

With this software it is possible to find information instantly. The full-text search tool allows the user to retrieve exactly what they need. Hypertext links can be used to simplify browsing. The navigation features such as, bookmarks and cross-documentation links, help the user move through numerous documents faster.

How PDF files work

Adobe Systems defines a PDF as a file format used to represent a document independent of the application software, hardware, and operating system used to create it. A PDF is a PostScript-language-based file format that describes documents that contain any combination of text, graphics, grayscale images, and color images in a resolution and device independent metafile format.

The PDF file format is not a programming language like PostScript. You cannot send a PDF file to a laser printer directly because the file format contains information that a PostScript RIP would not understand. The PDF does contain PostScript code, but the extra PDF data would inhibit the RIP from processing the document. A PDF file must be sent to a RIP through the Adobe Acrobat software application. The PDF is then converted into a PostScript language file and treated as any other PostScript document, either by downloading it directly to the RIP or by placing it into a page layout program such as QuarkXPress.

Differences between PostScript and PDF

• A PDF file may contain objects such as hypertext links that are useful only for interactive viewing.
• To simplify the processing of page descriptions, PDF provides no programming language constructs.
• PDF enforces a strictly defined file structure that allows an application to access parts of a document randomly.
• PDF files contain information such as font metrics, to ensure viewing fidelity.
• PDF requires files to be represented in ASCII, to enhance document portability.

To improve performance for interactive viewing, a PDF defines a more structured format than that used by most PostScript language programs. PDF also includes objects, such as annotations and hypertext links, that are not part of the page itself but are useful for interactive viewing.

Creating a PDF file

The two most common methods to create a PDF is either through the application software in which the document was created or from PostScript files made from the document. Many applications can produce a PDF file directly. The PDF Writer, available on both Apple Macintosh computers and computers running the Microsoft Windows environment, acts as a printer driver. The PDF Writer shows up as a printer in the Macintosh Chooser window. The user needs to choose that "printer" to create a PDF file. The user then "prints" their file to the PDF Writer and an electronic file is produced. This is similar to "print to disk." The PDF file that results is now platform independent; it can be viewed or printed from Macintosh, Windows, PC, and UNIX platforms.

For more complex documents that involve high resolution images and detailed illustrations, the PDF file must be created differently because of limitations of the system software. The application Acrobat Distiller was developed for this situation. Distiller was designed to produce PDF files from PostScript files that have been "printed to disk." The Distiller application accepts any PostScript page description, whether created by a program or hand-coded. The Distiller produces more efficient PDF files than PDF Writer for some application programs.

Viewing and editing PDF files

You can view a PDF file with Acrobat Exchange, Acrobat Reader, and Adobe Illustrator 5.5. Acrobat Reader is free, and it is available through almost all online services. Copies of the reader can be shared with others. Since PDF files can contain thousands of "pages" it is wiser to view these files in either Exchange or Reader. These two applications contain the interface that allows users to easily navigate through a PDF document.

If alterations need to be made to a PDF file it can be done in Adobe Illustrator and Macromind FreeHand. The file is somewhat editable in these applications. Document interchange is easier now that these two software packages can read and edit PDF files. The Associated Press in New York, which currently distributes graphics to 1,555 member newspapers in FreeHand format, switched to PDF in 1995, allowing newspapers to edit the files in either FreeHand or Illustrator.

Portability

PDFing a file makes it "portable" across computer platforms. A PDF file is a 7-bit ASCII file, and uses only the printable subset of the ASCII character set to describe documents—even those with images and special characters. As a result, PDF files are extremely portable across diverse hardware and operating system environments. Before PDF technology was invented, a major problem with sending documents to other users was the issue of font usage.

The individual on the receiving end needs those particular fonts used in the original document. In most cases, the receiver does not have those fonts and they would substitute other fonts for the missing ones. The result would be a document that no longer had the look and feel the sender intended. In some instances, this is more important than the actual message. To overcome this problem, some users send the fonts along with their documents. This becomes awkward when documents use many typefaces.

Acrobat provides a solution that makes a document independent of the fonts used to create it. A PDF file contains a font descriptor for each font used in a document. The font descriptor includes the font name, character metrics, and style information. This is the information needed to simulate missing fonts and is typically only 1–2 K per font. If a font used in a document is available on the computer where the document is viewed, it is used. If it is not available, a Multiple Master font is used to simulate on a character-by-character basis the weight and width of the original font, to maintain the overall *color* and formatting of the document.

When a PostScript file is created, the font outlines are included in the file. Distilling this file will ensure that pages are displayed with type characters in exact position, and the use of Multiple Master fonts will not be necessary. The only drawback to this is that the resulting PDF files will be somewhat larger. This works great for the Latin character set, but it will not work with pi or symbolic fonts. Symbols used other than Symbol or ITC Zapf Dingbats must be saved along with the PDF file. It is not possible to effectively simulate a symbolic or pi font.

PostScript allows users to design and create pages without concern about a particular output device. This device independence is the hallmark of PostScript and the key to success for desktop publishing. To produce any document, you should not have to think about where it is going, or what applications or which fonts are in use.

Document portability is a concern. Computer users have had to suffer from a lack of formatted text, loss of graphics and lack of proper fonts installed on particular computers that are used to view and print documents. Documents have been *somewhat* portable through the use of ASCII and RTF files, but content alone does not always convey the true message without formatting. There is no guarantee that a document created on one computer can be used or accessed by another. Documents can lose crucial data or be completely unintelligible unless the recipient has the same software and hardware as the author. Document exchange standards, such as ASCII or RTF, are useless for delivering richly formatted documents or graphics.

The look and feel of electronic documents must be maintained to make an electronic document portable. The PDF file format is a true portable document file format. PDF is a file format used to represent a document in a manner independent of the application software, hardware, and operating system used to create it. A PDF file contains the PDF document and other supporting viewing data.

Acrobat Distiller transforms PostScript files from PC, Macintosh, and UNIX systems into Portable Document Format (PDF) documents for viewing and printing in Acrobat Reader and Exchange. Distiller-created PDF documents retain all text, formatting, graphics and photographic images that the original documents contain.

Fonts

Electronic documents lose their format when viewed on other computers and computer systems. This is partly due to the lack of access to the exact font or fonts used to create the document. The portability of PDF documents overcomes this drawback. The PDF file has the font information imbedded.

The first version of this file format did not allow for font inclusion. The original Acrobat concept of using Multiple

Master font technology was unacceptable to advertisers with specific type styles. There had to be provisions for embedding fonts in the PDF file. Multiple Master technology simulates fonts that were used in a document if those fonts were not embedded. Acrobat and ATM now use the Multiple Master technology to simulate the font metrics (spacing, sizing, and other typographic characteristics).

Although font substitution preserves the overall look of your document, it doesn't always result in a document that is typographically what you were trying to achieve. This is a good practice to follow because of the difficulties that font substitution technology has with certain fonts and with certain typographic aspects of all fonts substituted. Embed your fonts—this is good Acrobat practice in general. Always embed display fonts—that is, unusual or fancy fonts. Engraved, script, and other non-standard typefaces will not be substituted correctly if your reader does not have the font you used.

PDF files are not totally dependent on the fonts that are embedded into your file, though. When required fonts exist on user systems, Acrobat renders documents using system fonts. To supplement system fonts, Acrobat includes and uses one symbol font and 13 basic PostScript fonts. Document fidelity is maintained when fonts are embedded in PDF files.

Compression

PDF files are portable files that have image and text compression applied. To minimize the size of PDF documents, the Distiller provides different data-compression methods for its three information classes. Those information classes are text and graphics, color and grayscale images, and bi-level (monochrome black-and-white) images. The compression of these files can be quite impressive. Starting from a large multi-megabyte PostScript file, the resulting PDF file could fit on a floppy disk.

Starting with a PostScript file of 31 Mb, you can distill a broadsheet page down to a PDF. With no compression options turned on, file size goes to 18 Mb. By using LZW compression on images, the file size goes down to 2.9 Mb, and with JPEG set for medium compression, the file size is 1.6

Mb. Compressed files are smaller and can be sent to the RIP and thus image quickly.

PDF and PostScript

Although PDF files require PostScript information to be created, the resulting PDF files are different from their PostScript counterparts. A PDF file is not a PostScript language program and cannot be directly interpreted by a PostScript interpreter. However, the page descriptions in a PDF file can be converted into a PostScript language program.

PostScript is considered a de facto standard and spurred the desktop publishing revolution. It is not a *portable* file format. PostScript was initially proposed by standards committees and association members as an interchange format for advertising and has been used on the Internet for the storage of documents. PostScript files tend to be very large, continuous, uncompressed and unstructured files with pages that may come out of order or may have resources that depend on previous pages. PDF, in its way, makes a lot more sense for data interchange—for advertisements, for example, because of its page independence—since each page is a separate unit, independent and compact.

Color

Adobe did not initially design Acrobat and its PDF documents to have extensive color quality or presentation capabilities. Users have pushed the format for publication ads and magazine quality transportation of documents. That meant that a PostScript file coming into the Distiller would lose some information—not vector shapes, typefaces, words or other letters—but information having to do with underlying PostScript commands, like screen operators. Upgrades to Acrobat and the PDF file format will affect high-end graphic arts use: color management, color space resources and function resources. "Supra" and "Bravo" will extend them.

Font usage

It is difficult to guarantee the intended look and feel of an electronic document when viewed on another computer. This is a concern to most users since not all computers use or have the same fonts used in various documents. PDF files can maintain the look of a document on other computers and computer platforms. The Acrobat PDF file format is compatible with font technologies, such as Type 1, TrueType and Multiple Master. Apple GX Fonts are not compatible.

Acrobat provides the option of *not* embedding fonts into PDF files. Choosing this option will reduce file size and font fidelity. This option should only be considered if the document is intended to be viewed only on the screen. Many PDF files retrieved from the World Wide Web are created in this way. The fonts that are viewed in these files are rendered and/or created by font substitution technology, which involves the use of two Multiple Master fonts, Adobe Sans and Adobe Serif. Adobe Acrobat uses these two fonts to create fonts that somewhat resemble the fonts used in documents. This substitution is unsatisfactory to the trained eye, and does not work in all situations. Faces such as a script, decorative, display, ornaments and expert series will not resemble the original fonts.

If fonts are important to the look and feel of the document, fonts must be embedded into the PDF file. This will ensure that the fonts shown on the screen and used to print the document are the original ones that were used to create the document.

Graphics

Many types of graphic file formats used in conjunction with desktop layout applications are usually compatible file formats to use within Acrobat PDF files. Raster file formats such as TIFF, EPS, EPS–DCS, EPS vector files and bit-mapped files are all compatible. All of these common file formats can sometimes be quite large. Acrobat Distiller can compress these files in a variety of ways so that the documents are smaller and more portable.

Users generally agree that a compression setting of either JPEG High, JPEG Medium Low, or LZW will produce acceptable compression results for four-color images. LZW compression produces acceptable results for grayscale images. Keep in mind that bit-map compression varies according to the image being compressed.

Page layout

Single- and double-page spreads can be converted into the PDF file format and viewed in the proper orientation. Editing PDF files cannot be easily accomplished. Other applications can read PDF files with PDF filters installed into those applications; but the file itself cannot be edited easily. Adobe Illustrator is one of the applications used to edit PDF files. When a PDF file is opened in Adobe Illustrator one can edit text and vector graphics. The text is still considered text and not outlines. Although the text is editable, in some cases the fonts do not transfer to the Adobe Illustrator format. This was initially due to the beta PDF plug-in that allowed Adobe Illustrator to read and save PDF files. The fonts that were converted to outlines were generally alright but these glyphs were no longer searchable text, but line art graphics.

Trap information created in QuarkXPress cannot be edited because the trap information does not convert correctly to a PDF file. Traps created in Adobe Illustrator *are* converted correctly into the PDF format and can be edited when put back into Adobe Illustrator. Vector graphics that are part of PDF files are easily editable when opened up in Adobe Illustrator. Raster images are more difficult to work with. In Adobe Illustrator, at present, there are no tools to edit raster images placed in Illustrator files. Opening up PDF files in Illustrator uses large amounts of disk space. It is probably best to go back to the originating program for such changes to the rasterized images.

Raster images

The raster images that are embedded in PDF files are decompressed and saved to disk. After editing is done, the resulting EPS file is enormous in size when compared to the original Illustrator file. A change from a 3 Mb original EPS to an edited 17 Mb EPS is not uncommon. These example file sizes are prior to conversion into a PDF. Editing PDF files that contain separated information must be treated one separation at a time since Adobe Illustrator cannot display more than one page at a time. Editing PDF files at this point in Acrobat's technological development is not a practical operation to perform in a production environment.

PDF conversion

PostScript and PDF are closely related since both have the same roots. PostScript is needed in order to produce PDF files for use in a prepress environment. The Acrobat Distiller strips out important imaging information concerning images contained in the resulting PDF file. Plug-ins are needed to add functionality back into Acrobat so that the PDF file can be used in a prepress environment. One such plug-in commonly used in Acrobat is "addPS." This allows users to have more control over how files are imaged on a film imagesetter. Making a PDF file is a relatively simple process. The long way involves saving your document in PostScript form and then opening it in Distiller. The easy way is to use the PPD printer driver, which does both things automatically.

Prepress issues

PDF files can contain color separation information by saving specific color information as individual PDF pages. This is the manner in which the color separation process must be done since there is no color separation engine in Acrobat Exchange, Reader or Distiller to perform it at output or when files are being distilled. Page layout applications used in conjunction with a PostScript driver can create a distillable PostScript file that can contain multiple pages of color information. For four-color printing, the PDF file contains four pages that correspond to the four color process inks. If the document has more than four colors, the PDF file would then contain more individual pages for the additional color information. Each color is given its own page in the PDF. For more colors there would be more pages.

The PDF file format has difficulties in translating trap information from some original applications. To apply traps in Adobe Illustrator one must convert the text into outlines. The application then no longer considers this information as text and treats it as a graphic element. Some of the high-end imaging capabilities of PostScript have been left out of the PDF file format.

Output

The Distiller strips out certain information in the process

of making PDF files. "addPS" adds PostScript information to a PDF document that describes what the screen frequency and angle should be for the PDF file being imaged. This plug-in is used to send PostScript files to a RIP allowing the user to change the default line screen and angle information of the imagesetter and RIP. The next file that is imaged at the image-setter would be imaged according to those specifications. A separate PostScript routine needs to be sent each time a new color is imaged because the angles need to be offset from each separation. If an imagesetter cannot set the screen frequency and screen angles without the help of a page layout application "addPS" must be used.

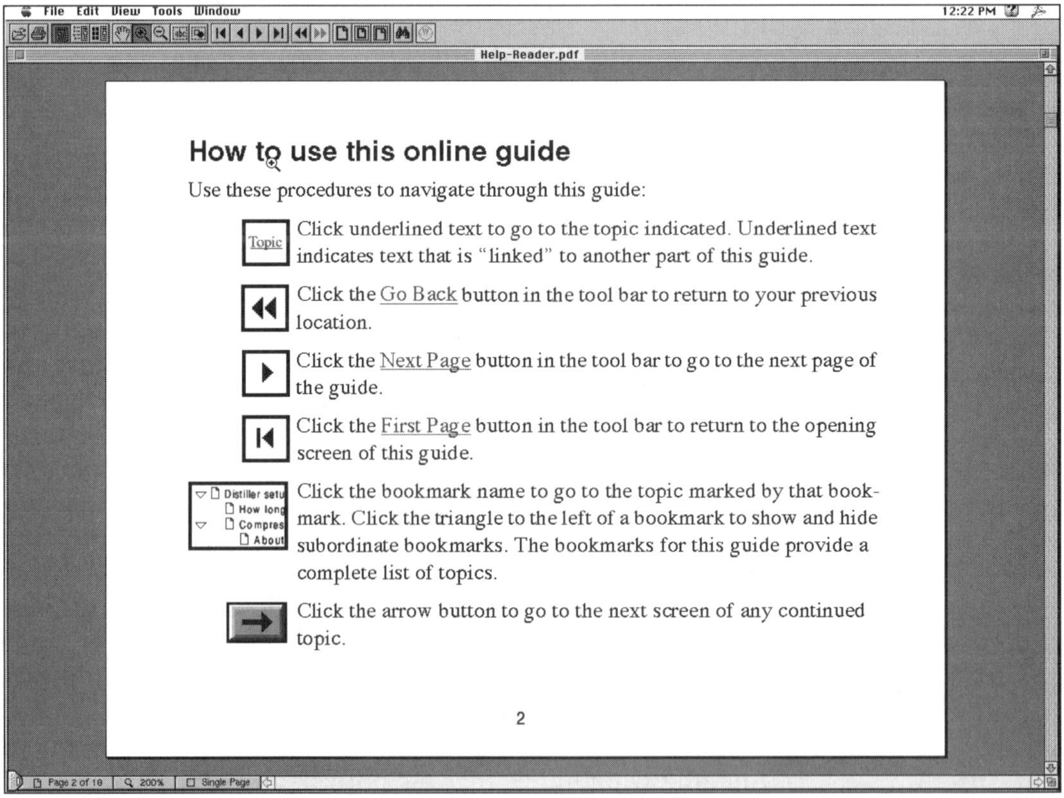

Tips on using Adobe Acrobat

Reader Window: A document will open in the reader window. Bookmarks and thumbnails can be displayed in the overview area to the left of the document. Below the menu bar is the tool bar, these tools are used for selecting and viewing documents. At the bottom left hand corner is the status bar. On the right hand side of the document is the vertical scroll bar, this enables the user to scroll up and down a document. There is also an independent scroll bar used to scroll through the bookmarks and thumbnails. In order to select a tool the user must click on the tool icon that they desire. If you want to hide or show the tool bar, select the Window Menu and choose Hide Tool Bar or Show Tool Bar. The following tools can be used as described.

File and **Print** icons are the first, going from left to right.

The **Page Only** button is used to close the overview area in the window. It only displays the page.

The **Bookmarks and Page** button is used to display bookmarks that were created in the document. To go to that location, just click on the bookmark's name.

The **Thumbnails and Page** button is used to display the thumbnail images of each page. To use, click on the thumbnail. Thumbnails are created in Exchange.

The **Hand** tool is used to move the document page when it does not fit within the window. To use, select the hand tool and then click and drag to move the page.

The **Zoom In (+)** and **Zoom Out (–)** tools are used to increase and decrease the page view size.

The **Select Text** and **Select Graphics** tools are next.

The Browser or page navigation arrows get you to the Start and End or Next and Previous pages.

The **Go Back** and **Go Forward** buttons retrace your steps in a document, moving to each view in the order visited. Go Back will return you to the original document after you click a link to another document.

Actual Size displays the page at 100 percent.

Fit Page scales the page to fit within the window.

Fit Width scales the page width to fill the width of the window. Pressing this button and the Option (Macintosh) or Control (Windows) key fills the window with only the visible text and graphics on the page.

The **Find** tool searches for part of a word, a complete word, or multiple words in a document.

Links

Hypertext links can be created in Adobe Acrobat Exchange. These links are used to connect parts of a document together. The hypertext links can link one PDF file to another PDF file. It is also possible to set up links between files and movie clips. To use the link, just position the pointer over the area and click. To get back to where you began just click *Go Back* on the tool bar.

Notes

To open notes that are used in a PDF file just double click on the note icon. When finished with the note, click in the box in the upper left hand corner.

Adobe extends its franchise

Adobe Systems Inc. has introduced a portable, highly-efficient imaging model code-named "Bravo" that is independent of the underlying operating system and designed to revolutionize the process of delivering graphically rich applications and content for the World Wide Web. Based on Adobe's industry-standard PostScript imaging model, "Bravo" will provide developers with the first integrated API (application programming interface) for displaying, printing and managing high-fidelity images, line art and text for Internet-based graphics. Adobe intends to establish "Bravo" as an Internet standard through key partnerships and by exploiting the technology within its own line of graphics, publishing, printing and Internet-based products.

JavaSoft, an operating company of Sun Microsystems Inc., intends to license the "Bravo" imaging model and incorporate it into a future implementation of the Java platform.

JavaSoft intends to license the Adobe "Bravo" imaging model for inclusion in the Java platform. "Bravo" will be included with Java as the first implementation of the collaboratively developed 2D specification in the JavaMedia family of APIs, which allow Java developers to produce highly interactive, extensible, dynamic network programs with rich media. Sun's success in licensing Java to key industry players such as Netscape, Oracle Corporation and Microsoft Corporation will help establish "Bravo" as an imaging standard for network computing. Netscape plans to support it in order to bring this new functionality to the over 20 million Netscape Navigator users.

"Bravo" will let developers to create a new class of highly graphical, platform independent applications that will run on personal and network computers. The Adobe imaging model promises to provide a graphical infrastructure that will enable developers to deliver high-fidelity, uncompromised graphics and provides access to advanced imaging capabilities such as true color images (16.8 million colors) with transparency, scalable PostScript line art and text that can be anti-aliased on-the-fly to eliminate "jaggies."

Text support in "Bravo" is not tied to any specific font technology—applications can interchangably work with Type 1, Type 1 multiple master, TrueType, and other type technologies including the recently announced "OpenType" specification. In addition, "Bravo's" printing architecture is optimized for professional graphics and publishing applications, directly generating PostScript language commands, improving the quality and speed of printing for complex graphics. Because Bravo is derived from the device independent PostScript imaging model, users will be ensured high-fidelity experience between the screen and the printed page.

Adobe exploits "Bravo" in upcoming applications

Already a proven technology, Adobe "Bravo" is broadly used today in a number of existing Adobe applications. Adobe intends to further incorporate the technology in future versions of its products as new releases are developed. Bravo will also be the basis for entirely new desktop and Web applications—the first of which will be a new interactive authoring tool from Adobe, code-named "Vertigo." Adobe has demonstrated more than a dozen of its Web publishing products and technologies including a new interactive authoring technology, code-named "Vertigo", and a derivative of Adobe Persuasion, code-named "Adobe Web Presenter", that enables users to create presentations for the Internet.

Adobe's plan for improving the Internet experience

• Building the best creation tools for creating compelling, high quality, dynamic information at every level of the Internet.
• Providing the best way to deliver and print information beginning with PrintMill, a printing solution that enables users to print to any printing device across a network.
• Driving and supporting Internet standards, including existing standards such as HTML, GIF and Java. Adobe intends to raise the quality of Internet graphics through the following:
- Type Extensions that improve browser quality by maintaining native fonts, size and formatting
- Embedded Portable Document Format (EPDF) capabilities that enables users to place rich graphics in any Web document or application
- Adobe "Bravo"—a portable imaging model that sets a new standard for creating interactive, graphically rich applications and content for the Web.

6.

SGML AND HTML

In 1979, Charles Goldfarb, working for IBM, submitted to the American National Standards Institute a text-based computer language that he called GML, or Generalized Markup Language. Over the years, it has been tweaked and become part of the public domain as Standard GML, better known as SGML.

Goldfarb joined IBM in 1967, after working in a law firm for several years, to create a document management system for the legal profession. The problem that he faced was that to get legal briefs from an IBM mainframe onto paper, complex assembler commands were needed. While not a monumental problem in itself, when you take into consideration that three programs were needed to perform the legal research, and each used different program codes, things became quite difficult. And that is the reason why Goldfarb went about developing GML, from which SGML was born.

So what exactly is SGML? Simply put, it is a rigorous, non-proprietary, descriptive markup language. SGML provides an internationally recognized, non-proprietary language for writing your own markup schemes. Markup is the text that is added to the data of your files in order to convey particular information about that data. In a word processor, the markup is the proprietary codes that the software inserts into your text files to indicate which words should be printed in a certain font, which paragraphs should be centered, where page breaks occur, etc.…

SGML relies on the principles of descriptive markup where the markup is used to indicate the nature, function or content of the data in a file, rather than saying how that data should be processed. SGML can be used to encode semantics rather than syntax. Using SGML, a heading would be identified as a *heading* rather than a piece of text that has to be printed or displayed in *20 point Times Bold*.

The markup schemes that you write using SGML declare a set of rules which unambiguously state how the data must be marked up in order to be correctly structured. SGML-aware software can ensure that any markup in a file conforms to the appropriate set of rules thereby guaranteeing that the data in the file will be structured in a known way. If the set of rules you are working with declares that text labeled as a 'sub-section' can only occur within text labeled as a 'section', SGML-aware software will ensure the rule is obeyed during text creation and editing.

It used to be that the text standard for files of different programs and platforms was ASCII. However, ASCII in its earlier standard 7-bit form only allowed for 128 glyphs and controls. There was also very little formatting of text in different layouts, fonts, and sizes. Standard character sets (like ASCII) were all the standards that one needed to exchange electronic documents. Despite its versatility, ASCII has many limitations. In the post-Macintosh era, in the world of computer assisted publishing, we need new standards with support for layout, typefaces, graphics, sounds, movies, hyperlinks and everything in between. Numerous incompatible standards for electronic document exchange exist and all those standards emphasize common appearance of documents: recipient and sender must see almost the same thing. SGML, on the contrary, concentrates on document structure and totally ignores visual appearance.

It is this concentration on structure that sets SGML apart from formats such as RTF, PostScript and Adobe Acrobat. This makes it a very powerful and flexible contender in many fields from multimedia to traditional documents.

One of the biggest misconceptions about SGML is that it is a standard document structure. However it isn't and the reason why it isn't is that no one standard structure could be useful to all authors. The structure that may be perfect for books, might not ever fit the needs of letters and timetables. So SGML is an enabling standard, and it is up to the Document Type Definition, or DTD, to define what the tags (or codes) are and their relationships with each other. DTDs are also often referred to as an SGML application. SGML does not set the specific structure of the DTD; it is up to industry groups and committees to set the standards for the DTDs. An example of a DTD is MIL-M-28001B, a specification required by the Department of Defense for submission of technical manuals. And while not strictly a DTD, the Hypertext Markup Language, HTML, is often referred to as one.

HTML is an application of SGML, not a subset of SGML. More in-depth explanations are not so easy to come by, because HTML was not originally written as an SGML application. Only now is the HTML language being given the formal rules required by the SGML standard.

The reason HTML is commonly known is that it is the format used by the World Wide Web on the Internet. The World Wide Web, also known as the Web, was created by the European Council for Nuclear Research (CERN) so that they could use the Internet to share information with other scientists throughout the world. The protocol they created was HTML; a deliberately simple set of tags that carried basic information such as type style and hypertext links. The reason they used so few and such simple tags was so that the programs needed to interact with the Web could be small and simple, and so that use of bandwidth was kept to a minimum. In its early days, the Web was quite small, and text only, thus there was no real need for complex tags and commands. However, quite recently, the Web has exploded because the first graphic-based Web Browser was introduced by the National Center for Supercomputing Applications called Mosaic. With Mosaic, it was now possible to create much more complex Web sites with different fonts and varying point sizes, graphics, and numerous other forms of multimedia such as sound, movies, and computer programs.

Now that the scope of the Web has expanded far beyond its original use, the current form of HTML is woefully inadequate. In December of 1994, HTML 2.0 was formally proposed and when HTML 2.0 is finally adopted, HTML will finally be a true DTD. HTML 3.0 is also in the works and should be drafted by the end of 1995 to early 1996. This version will allow for tables and style sheets, that are already being used in some extended HTML viewers such as

Netscape 2.0. SGML's (and HTML's) greatest strength is in structure, not layout.

With the Web becoming a major area in online publishing, many publishers are turning, quite ironically, back to incompatible proprietary formats such as Adobe Acrobat. The reasoning behind this is so that when the viewer receives a document from the sender, it looks just like it did, when the designer made it. It is hoped that future versions of HTML will have more complex tags so that this won't be needed.

The development and explosive growth of the World Wide Web has necessitated a need to develop new tools and techniques designed to harness the technology involved in Internet publishing. Currently, a World Wide Web publisher is faced with a myriad of hardware and software considerations for the development and maintenance of a site.

The largest document

The World Wide Web is the world's largest "document"—with thousands and thousands of text, graphics, audio and video files interlinked throughout the world on connected servers. Developed in 1991 at CERN, the European Particle Physics Laboratory in Geneva, Switzerland, it was originally conceived as a hypertext exchange system to allow scientists to share their findings with other scientists over the Internet. The World Wide Web was met with great enthusiasm and quickly grew beyond the bounds of CERN. Computer enthusiasts of all kinds started developing World Wide Web applications and promoting the World Wide Web across the Internet. It did not take long before all types of hypermedia became available on the World Wide Web for Internet enthusiasts to explore. The World Wide Web's use and popularity began to blossom, suggesting the future of hypermedia would grow.

Unfortunately, at that point, the World Wide Web was still difficult for anyone but Internet professionals to navigate. This changed in June of 1993 when the National Center for Supercomputing Applications (NCSA) introduced a simple and powerful graphical user interface for World Wide Web navigating. The software is called Mosaic, which can be retrieved from the Internet for free, and is currently revolutionizing the World Wide Web and bringing it to the masses. In June of 1993 a software robot was sent through the World Wide Web to count how many sites existed. The robot found around one hundred such sites. In May of 1994, a similar robot logged onto 3,800 sites. This significant increase in new sites was due almost entirely to Mosaic.

The structure of the World Wide Web allows for this great influx of new sites and the millions of World Wide Web users visiting these sites. The World Wide Web is based on the Hypertext Transport Protocol (HTTP), a protocol designed for sharing documents via a client server model. Hypertext Markup Language (HTML) is simply the language that all World Wide Web documents speak. It is a set of tags that tell the navigational software how to display text, hypertext links, images, and movie clips. HTML was purposely kept simple; there are only a few tags for text markup, allowing navigational software to be minimal and thus inexpensive to write. Anyone can download Mosaic for free; and it was actually written by students.

The simple tagging schemes of HTML allow novice World Wide Web users to create their own documents. These homemade documents or home pages, can now be hyperlinked to any other home pages residing on a World Wide Web server via their Uniform Resource Locator (URL). A URL is a system that the World Wide Web uses to standardize the addresses of various Internet services. A URL "expresses the address of a resource and the method by which it can be accessed." In short, a URL provides location information. A typical URL looks like this:

```
http://www.ncsa.uiuc.edu/SDG/Software/Mosaic/M
osaicHome.html
```

Within the URL, text are components that specify the protocol, server, and pathname of an item. The protocol is followed by a colon (http:), the server is preceded by two slashes (//www.ncsa.uiuc.edu), and each segment of the pathname is preceded by a single slash:

```
/SDG/Software/Mosaic/MosaicHome.html
```

The first component, the protocol, identifies a manner for interpreting computer information. Many Internet pages use HTTP (short for HyperText Transfer Protocol). Other common protocols you might come across include file (also known as ftp, short for File Transfer Protocol), news (the protocol used by Usenet news groups), and gopher (an alternative transfer protocol).

The second component, the server, identifies the computer system that stores the information you seek. Each server on the Internet has a unique address name whose text refers to the organization maintaining the server.

The last component, the pathname, identifies the location of an item on the server. For example, a pathname might identify a page by specifying the name of the file comprising the page (/MosaicHome.html) as well as the name of one or more folders that store a file (/Mosaic).

Since Mosaic is such a "bare-bones" software package, supplemental programs are needed to achieve full benefit of the World Wide Web's hypermedia capabilities. These programs reside on one's hard disk and are referred to as Helper programs. This scheme allows Mosaic to keep its focus on its business, while letting other sophisticated viewing and listening applications handle the media presentations. Depending on the platform running, it obviously dictates which supplemental programs are used. Typical programs running on a Macintosh are SimplePlayer for viewing QuickTime clips, Sparkle for viewing MPEG clips, SoundMachine for AU sounds, and JPEGView for viewing GIF/JPEG color and monochrome images.

HTML offers great flexibility in its tagging schemes. Tags are guidelines for how a software navigator should display the material. Some tags are required by all HTML software navigators, while other tags are optional. All World Wide Web software navigators are different. Some, like Mosaic, are graphical, while others are text based. HTML is only a set of guidelines for displaying the documents—not everyone has to have the same capabilities to access World Wide Web information, such as text-only users. Once the HTML document arrives at its destination, the software navigator can use whatever tags are appropriate to it and ignore the rest. For instance, Lynx, another popular World Wide Web browser, supports text only. It therefore ignores the HTML style information that's not relevant to it. Mosaic, on the other hand, presents documents in all their multimedia glory. This flexibility allows you to change the way World Wide Web documents look on your computer system.

The overwhelming success of Mosaic brought an onslaught of new software navigators, all vying for their share of the marketplace. Currently, Netscape Communication's World Wide Web Browser, called Netscape Navigator, has overtaken NCSA Mosaic as the de facto World Wide Web navigational software. Netscape Navigator supports later releases of HTML, therefore giving the designer more control of the style of their document.

Hypermedia is a major attribute of the Internet, but all file retrieval capabilities are at the mercy of bandwidth. Direct or dedicated connections to the Internet are suggested to take as much advantage of the World Wide Web's media delivery capabilities. Using SLIP or PPP dial-up accounts to retrieve QuickTime or MPEG video clips may be frustrating. SLIP/PPP connections use high-speed modems (28,800 baud) and special software to connect a computer to the Internet.

Reasons for interest

The pen is no longer the only implement mightier than the sword, and freedom of the press no longer necessitates a press. The mighty pen and inky printing press are being challenged by a quirky language called HTML (Hypertext Markup Language) and a paperless and inkless press known as the World Wide Web.

The printed medium is a wonderful source for disseminating information and will assuredly exist forever. However, more glamorous, cheaper, quicker, and environmentally sound methods of communicating the same information are hitting the mainstream. One does not have to be a printer. One might be an electronic publisher, an information gatherer, converter, transformer, and deliverer. This outlook creates a wealth of new opportunities and challenges. Repackaging information in new ways creates opportunities, and develops new markets, and new audiences.

Print, interactive multimedia, and online publishing are pockets of opportunity, and now relatively saturated. However, collectively, these media preview the future of information delivery. The opportunities are endless, limited only by the imagination of human endeavor, and available bandwidth. Currently, the World Wide Web allows all these individual information technologies to merge, thus creating a new paradigm for information delivery.

The effective integration of digital media elements such as graphics, audio, and video into a World Wide Web site can be just as important as informative and well-organized text. When done properly, the incorporation of multimedia into a site can substantially increase its utility as well as its aesthetic appeal. However, poorly implemented multimedia can just as easily diminish a site's usefulness. When dealing with multimedia elements, a number of issues must be carefully considered. In stark contrast to ink on paper, publishing on the World Wide Web can be described as instantaneous communication, distributed worldwide to millions of readers in content-rich, colorful, highly interactive format, produced for a unit cost too cheap to meter. There is only one problem: they have to know how to find it.

Data structure

The learning curve associated with setting up and maintaining a WWW site is additive. As a WWW site administrator or Webmaster, becomes comfortable with the site's basic functionality, more advanced features can be added. Implementing these features requires greater levels of technical expertise. Implementing mixed-media elements into a World Wide Web document increases the overall file size of that document. The more elements that are included, the higher the file size. This is the single largest issue confronting online information providers to date. The technology allows for inclusion of video, sound, and large graphics, but most connections are ill equipped to download the data within a reasonable time frame.

Certain navigational software tackles the large graphics issue head-on. Users can specify parameters whether to allow the graphics to be viewed or not. If a user chooses not to show the graphics, a small icon appears notifying that a graphic exists. This significantly reduces retrieval time.

Audio and video files are relatively large, thus exacerbating the file size issue. Content providers who implement audio and video files within a document must seriously consider the end viewer when deciding the size of such elements. It is a general rule that potential downloaders are warned of any files over 100 K.

Another important consideration for graphics files is their physical size (height and width). Although it is directly related to data size, it is important to keep the physical size within reasonable limits so it looks acceptable on most viewers. Similarly, it is important to use reasonable length video and audio clips (less than one minute). Various software packages offer techniques on how to address this problem.

Data formats

World Wide Web publishers create their multimedia elements across the gamut of computer platforms utilizing hundreds of software packages. The resulting effect is a myriad of possible file formats facing the World Wide Web publisher. Also, not all navigational software and supplemental programs are suited to handle the different file formats. Certain standards exist that attempt to tackle this issue.

Text-only users

Many World Wide Web users still use text-only navigational software. In fact, until two years ago, text-only navigational software was all that existed. It is frustrating for a text-only World Wide Web user to try to access information, only to find out that the particular site is only formatted for graphical navigational software. These users are often ignored, reducing the value of a World Wide Web site. Strategies exist on how to best orient a World Wide Web site for all types of navigational software.

Although the World Wide Web has great support for graphics and hypertext links, it needs better security to enable financial transactions to take place safely. It has cried for a richer markup language to handle tables and style sheets. And it has required greater efficiency in expediting transactions. Although the Web is touted as the next great information delivery vehicle, there are still many shortcomings that must be solved before it is a mainstream approach.

HTML programming

Once the information is compiled, it has to be converted into HTML documents so the Web browser can interpret them. The Hypertext Markup Language is designed to specify the logical organization of a document, with important hypertext extensions. It is not a WYSIWYG word processor such as Microsoft Word or WordPerfect. This is because the same document may be viewed by many different browsers, of very different abilities. For example, HTML allows you to mark titles or paragraph marks, and then leaves the interpretation of these marked elements up to the browser. One browser may indent the beginning of a paragraph, while another may only leave a blank line.

HTML instructions are called elements. These can be divided into two broad categories—those that define how the body of the document is to be displayed by the browser, and those that define information about the document, such as the title or relationships to other documents. Elements are denoted by the tag <element_name>. This is simply the element name surrounded by left and right angle brackets. Most elements mark blocks of the document for a particular purpose or for formatting: the above <element_name> tag marks the beginning of such a section.

HTML documents are structured into two parts, the head, and the body. The head contains information about the document that is not generally displayed with the document, such as its title. The body contains the body of the text, and is where the document material is to be displayed.

Style issues

HTML is based on SGML (the Standard Generalized Markup Language), which is used to describe the general structure of various kinds of documents. It is not a page description language like PostScript, nor is it a language that can be generated by a page layout program. The focus of HTML is the content of the document, not its appearance.

With a few minor exceptions, HTML does not describe the appearance or layout of a document. The designer of a Web page has no control over line length, typeface, point size, or color. These characteristics are specified by the end user according to their Web browser, personal preference, or system limitations.

The obvious disadvantage of this approach is that the designer has limited control over how the document will be viewed. This frustrating situation takes some getting used to. Designers usually come from disciplines where they control the final output, and the thought of someone else controlling the final look of a document is an awkward one.

Working in a text-only markup language, with little control over the appearance of a document may seem frustratingly archaic compared with today's technology. But for the kind of environment that the Web provides, HTML does have advantages over other forms of document publishing languages. For example, each HTML document is small, so it can be transferred over the Internet as fast as possible. Typeface or format information is not within the document and thus allowing for faster loading and displaying. Also, HTML documents are device independent and can be displayed on any platform. This is absolutely necessary considering that all types of computer platforms access information off the Internet. All that is needed is a Web browser that can interpret HTML for that platform.

Inline versus external images

The IMG element allows an image to be inserted within an HTML document. This is known as an inline image whose function is to allow graphics to be included within a document and presented with the text. The opposite is an external image which has hypertext links that retrieve and display images in a separate viewing window. Both methods allow a user to display an image, and when used appropriately, can enhance a Web presentation considerably.

There are two methods for presenting images. Interlaced GIFs appear first with poor resolution and then improve in resolution until the entire image has arrived, as opposed to images arriving linearly from the top row to the bottom row.

The interlaced approach is great to get a quick idea of what the entire image will look like while waiting for the rest. Some browsers do not support progressive display as the image is downloaded, but non-progressive-display Web browsers will still display interlaced GIFs once they have arrived in their entirety.

External images

The only inline image file format that all browsers can view is GIF. Linking to external images gives more flexibility in what image file format can be used.

A common practice with Web pages is to provide a very small GIF image (a "thumbnail") inline on the page itself, and then link that image to its larger counterpart. This has two major advantages over including the entire image inline: It keeps the size of the Web page small, so that page can be downloaded quickly and it lets the reader get a feel for the image and decide whether it it suitable to download the whole image. This flexibility is an advantage when providing information to a large audience with a myriad of different system setups and requirements.

Sound

Sound data is a digital representation of an analog signal, which is typically represented as a continuous waveform. When sound is digitally recorded, samples of wave form are captured at fixed intervals. The more samples that are taken, the more information stored for each sample, the higher quality of the sound. Sample rate measured in kilohertz (KHz) describes the quality of the sound. Common sample rates are 11 KHz, 22 KHz and 44 KHz. Sample size, usually in 8-bits or 16 bits, is the amount of information stored for each sample. For example, audio CD has 44 KHz and 16 bits sample size, which is very high quality sound.

8000 sample 8 bit .au files are currently the best choice for putting sound files on the Web. The reasons for this choice are: this format can be supported by most of the sound-capable machines on the Internet. Secondly, this format produces a fairly small file that requires little bandwidth to transfer. AU files are of only barely acceptable quality, as the 8-bit sampling causes them to sound comparable to transmission over a telephone. There are other sound formats supported by Web browsers. It is important to note that most sound formats are specific to a certain platform.

Video

The four variables which affect the size of a video file are the length of the clip, the size of the frame, the color depth (256 colors, thousands, millions, etc.), and the amount of compression applied. Manipulation of all these settings determines what combination produces acceptable quality files with the minimum size.

Like other files, movie files can be identified by their file extensions. There are only few movie file formats that can be viewed from the Internet, which are international standard file formats for multimedia. They are MPEG and QuickTime.

MPEG is a very popular movie file format for PCs and stands for Moving Pictures Expert Group. The members of this group come from more than 70 companies and institutions worldwide including SONY, Philips, Matsushita and Apple. They meet under the International Standard Organization (ISO) to generate digital video standards for Compact Discs, Cable TV, Direct Satellite Broadcast and high-definition television.

QuickTime is an ISO standard for digital media. It was originally created by Apple Computer Inc. and used in the Macintosh. It brings audio, animation, video, and interactive capabilities to personal computers and consumer devices. QuickTime movies are real movies. This standard is more mature than the MPEG standard. In December, 1993, Apple announced that it had begun demonstrating technology that will make future television and multimedia devices more compelling, interactive, and useful for people. Specifically, Apple demonstrated the integration of MPEG technology into applications using QuickTime technology. QuickTime for Windows is available for customers who use Microsoft's Windows operating system.

QuickTime movies have file extension ".qt" and ".mov." .Mov files can be played on both Macintosh and PCs, while .qt extensions can only be viewed on a Macintosh.

Server issues

The chief hardware component of a Web site is the host computer, which houses the site's content. The software program that runs on the host is called a Web server. It interprets incoming requests and returns the appropriate documents. Server programs are available for most major operating systems, including Unix, VMS, VM, Macintosh System 7, Windows 95, and Windows NT.

The major considerations in choosing a WWW host are platform stability, performance, and amount of RAM (random access memory). Platform stability is most important. If the operating system routinely crashes under normal loads, the Web site often will be inaccessible.

In general, Unix computers have the most stable and robust operating systems, but their stability can be outweighed by their cost and technical requirements. Macintosh and Windows machines are acceptable for handling light loads (with peaks of several hundred hits per hour), but are not recommended for higher loads

Most of today's business hardware—including 486-based PCs, 68040-based Macs, and nearly every available Unix workstation—can handle the processor and I/O (input/output) demands of all but the most heavily trafficked Web servers. However, the host computer's physical RAM can cause a bottleneck as the usage load increases.

A basic WWW platform should have at least 8 Mb RAM, and high-load computers should include 64 Mb RAM. To serve heavy loads, the host computer should be a dedicated WWW machine—one that is used exclusively for the Web site. Under low loads a host can be used for tasks in addition to its Web server duties.

HTTP

HTTP (Hypertext Transport Protocol) is the primary protocol used to distribute information within the World Wide Web. It is a relatively simple, highly flexible protocol used to deliver information across the Internet network of networks. HTTP defines a simple transaction, consisting of the following four parts, to deliver requested information from a server to a client:

- *The client must first establish a connection to the server.*
- *The client then issues a request to the server specifying a particular document to be retrieved.*
- *The server sends a response containing a status code and the text of the document if it is available.*
- *Either the client or the server then disconnects.*

One main goal of HTTP was to provide a simple algorithm that would enable fast response times. To achieve this goal, HTTP was defined as being a "stateless" protocol—one that does not retain any information about a connection from request to request.

HTTP is also a "connectionless" protocol—limited to one request per connection. Unlike other protocols, such as FTP, the connection between server and client is broken after each request is made. This means that every time a client wants to fetch a document, it establishes a new connection to the HTTP server.

This is one of the main reasons why it takes so long to load HTML pages with many inline graphics. For each graphic, a separate connection is established and information is requested. While establishing a connection is not generally time consuming, it can seriously affect performance for distant or heavily loaded sites.

Some clients, like NCSA Mosaic, wait until a connection is closed before they open their next connection. However, many newer clients, such as Netscape, open multiple connections and receive documents in parallel. Unless bandwidth is the bottleneck in retrieving documents, this behavior results in significant time savings when accessing sites loaded with inline images.

Data size

The amount of data that has to be transported for image or video files is the most important concern for WWW publishing. This is especially important because of the large number of users on 56 Kbps or slower lines. For them, it can take minutes to receive even medium-sized images, sounds, or video files. Browsers are beginning to cater more to users on slow lines by providing extra functionality. Most browsers let the user delay or completely turn off image-loading functions. Several second-generation browsers request images simultaneously, or in parallel, which significantly speeds the retrieval process. However, while these features help, content providers must carefully watch the size of the images.

There are a number of techniques for reducing the data size of images. Although the most obvious way may seem to reduce the image's total physical size, this does not necessarily yield substantial data-size reductions. An important issue to keep in mind when generating images is that different data formats compress images differently. For a particular image, the JPEG format may be significantly smaller than the GIF format, or vice versa.

Although it may not seem apparent, reducing the number of colors in an image can produce substantial savings in data size. By halving the total number of colors, the file size can be reduced by 10% to 30%. Luckily, the change in the number of colors usually is barely detectable to users. Because most browsers already limit the number of colors seen by users (typically 25 to 50 colors), reducing the number of colors makes an image more in accord with the browser display and speeds up loading.

Physical size

Another important consideration is an image's physical size (height and width). Although it is directly related to data size, it is important to keep the physical size within reasonable limits so it looks acceptable on most viewers.

To fit completely within the windows of most Web browsers, images should not be wider than six inches. All images to be served on the Web should fit within this limit. Most image-editing programs can perform these functions.

One popular and useful way to reduce the size of the graphics on a page is to use thumbnail images as links to the main images. Thumbnails are small versions (approximately one inch square) of images. Besides being small (2 K to 5 K), they can save time by letting users decide whether or not to retrieve the full image.

Text-only users

Although a large number of Web users run graphical

interfaces, text-based Web clients such as Lynx continue to be popular. Many Web users also use their graphical browsers in text-only mode to speed document retrieval. These users are often ignored, reducing the value of a Web site.

The most frequent mistake made by content providers is not using functional ALT tags for inline images. An ALT tag is embedded within an HTML document, so non-graphical browsers can display something other than the default [IMAGE] label. This means that a site viewed with Lynx will contain a large number of holes with [IMAGE] labels. Perhaps even worse are the ALT tags that describe an image. Reading "a picture of our corporate logo" instead of viewing the logo does little for users. Browsers such as Lynx will insert the text directly into the Web document, which can confuse readers.

ALT tags should be functional, not descriptive. The ALT tag for a bullet icon should be "o" not "Bullet." A graphic used for a horizontal line should be a series of dashes ("-"), not the phrase "horizontal line." Always keep users with text-based clients in mind when writing HTML. This requires little effort and will greatly improve the utility and appearance of the site for those users.

The Web's format and the opportunity to reach millions of potential customers is creating new types of publishing ventures in true "gold-rush" fashion. The Web is being touted as the fourth medium, and some suggest it will have as great an impact on society as print, radio and television. The growth of the Web is explosive and will assuredly continue to grow as more content becomes available.

Upon completion of this study, the author remains skeptical whether the World Wide Web is the medium of the future. It has, however, created a trend which will forever reshape the publishing world and the way information seekers receive their data. Publishing will change from a commodity based market where prices are based upon cost, and shift to a service market where prices are based upon the value of the information. Each reader requiring selected information tailored to their specific choice and paying for what they select—not paying for an entire magazine or newspaper in order to read only one article. The future of information dissemination is electronic, interactive and selective. Whether the delivery mechanism will be the World Wide Web remains to be seen.

Currently, the Web is the most advanced information system deployed on the Internet, and is equipped to embrace many future advances in technology, including new networks, protocols, and data formats. The most prominent shortcoming of the Web is its inability to securely transfer private data. Once secure transactions can take place, the Internet will become a mass means of both information dissemination and commerce.

Utilizing the World Wide Web as a publishing vehicle is at the mercy of available bandwidth. To take full advantage of the Web's hypermedia capabilities requires a considerable amount of bandwidth. Continued research into data compression can alleviate a host of data transfer issues currently limiting the functionality of the Web. With data traveling at faster speeds, designers of Web pages have more flexibility in the information they offer. It is the author's opinion that increased bandwidth results in more dynamic Web documents. So get a really fast modem.

Basic Structure of an HTML Document

HTML documents are divided into two sections: a *head*, and a *body*. The *head* contains information about the document itself, such as the title, indexing information, and ownership. The *body* consists of the document itself, the images, text, and hotlinks to other documents and files displayed by the Web browser.

Head Section

```
<HTML>
<HEAD>
<TITLE>Sample Web Page</TITLE>
</HEAD>
```

The head section is opened and closed by the <HEAD> and </HEAD> tags. Information belonging in the head section should be placed between these two tags. The tags in the example above form a minimal head section, and should be included at the beginning of every HTML document.

Body Section

```
</HEAD>
<BODY>
<H1>Introduction</H1>
```

The body section is opened and closed by the <BODY> and </BODY> tags. The text and images of the document itself, to be displayed by the Web browser, should be placed within these two tags.

The opening tag should be placed immediately after the closing head tag (</HEAD>), and before any images or text included in the document itself. A matching </BODY> must be placed at the foot of the page. This tag ends the body section, and should be placed at the end of the document, after any images or text included in it.

Tags Ending a Page

```
</BODY>
</HTML>
```

</BODY> ends the portion of the page containing the document itself, and should be placed after any text or images that will be displayed by the Web browser. </HTML> ends the page itself. The head and body sections are enclosed within the opening <HTML> and closing </HTML> tags.

Creating a Basic HTML Document

Open your ASCII text editor of choice and type in the following code. Do not worry what <H1> means, you will learn what it is in the next chapter. Once you are done typing, save the document with the appropriate .html (.htm for DOS) extension; for example, sample.html.

```
<HTML>
<HEAD>
<TITLE>Exercise 1</TITLE>
</HEAD>
<BODY>
<H1>Introduction</H1>
</BODY>
</HTML>
```

Open up your Web browser of choice and select the function which opens local files (**File/Open File** in Netscape). Then, open the HTML file you created. Remember, you must save the file first in your text editor, then open it. For each iteration, you can also use the reload function of the browser. The graphical representation displayed by the Web browser should look something like below. If it doesn't, look over your code and fix any mistakes.

What are Headings?

There are six levels of headings. The larger the number, the smaller the size of the resulting text (using a graphical browser). Examples of all six levels are provided below, followed by the tags that produce them. Headings are used to logically organize a page into separate sections for easier comprehension. Lower level headings in turn subdivide each section. Headings automatically create space between them and any text preceding or following them.

<H1></H1> Level One Heading

This is the highest level heading. It is often used at the beginning of a document.

<H2> . . . </H2> Level Two Heading
<H3> . . . </H3> Level Three Heading
<H4> . . . </H4> Level Four Heading
<H5> . . . </H5> Level Five Heading
<H6> . . . </H6> Level Six Heading

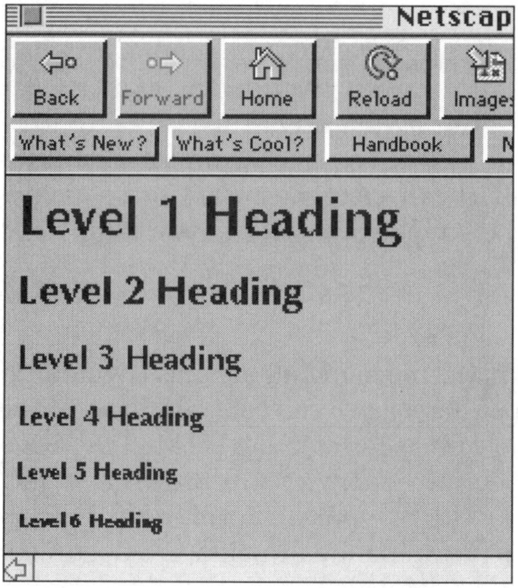


```
<HTML>
<HEAD>
<TITLE>Headings</TITLE>
</HEAD>
<BODY>
<H1>Level 1 Heading</H1>
<H2>Level 2 Heading</H2>
<H3>Level 3 Heading</H3>
<H4>Level 4 Heading</H4>
<H5>Level 5 Heading</H5>
<H6>Level 6 Heading</H6>
</BODY>
</HTML>
```

Controlling Text Spacing

HTML leaves the details of how to display a document (spacing, font selection and size, etc.) up to the program viewing the document. White space (such as tabs, spaces, and carriage returns or line feeds) is ignored. Therefore, specific tags marking paragraph and line breaks must be inserted into HTML documents.

Paragraph Tags

The <P> tag starts a paragraph, and the </P> tag ends it. These tags create a space between any images or text preceding the enclosed text (some browsers also create a space after the closing paragraph tag).

```
<P>
This is an example.
</P>
```

Line Break Tag

The line break tag causes any text following it to begin on the following line.

```
This is<BR>
an example.
```

Horizontal Rule

The <HR> element is used to draw a horizontal dividing line completely across the browser window, no matter what width the window is set to by the user. This can be to logically separate blocks of text, or to separate an icon list from the body of the text. The HR element is said to be *empty* which means there is no copy to apply it to—you don't need a </HR> to close it.

Creating an HTML document

Create a simple HTML file using all the tags covered up to this point. You should have a comfortable feel for the basic structure and tagging schemes for creating HTML files. Save your document and view it under a Web browser. If you have more than one browser on your platform or have access to other platforms, it is recommended to view the document under each. Each browser displays HTML slightly differently and the changes may be surprising.

```
<HTML>
<HEAD>
<TITLE>Exercise 2</TITLE>
</HEAD>
<BODY>
<H1>A Primitives Portfolio</H1>
<H3>By Malcolm Jones Jr.</H3>
<HR>
<P>

The one sure thing about Bob Dylan is that there
is no sure thing. In a musical career stretching
over more than three decades, he has proven time
and again that he owns the most bottomless bag
of tricks in the business...

</P>
<HR>
From Newsweek<br>
March 20, 1995
</BODY>
</HTML>
```

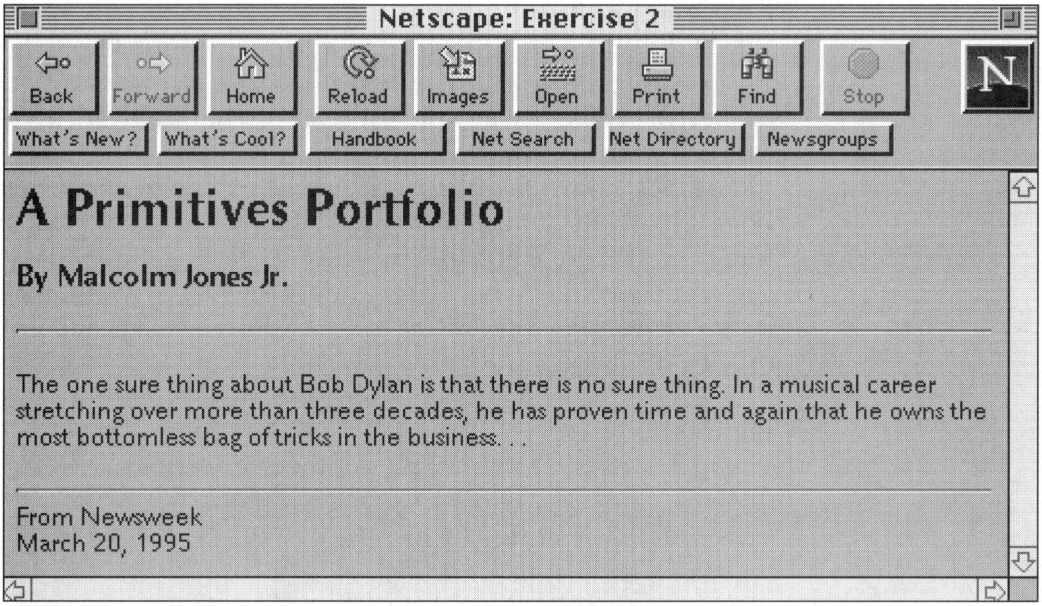

The output in Netscape

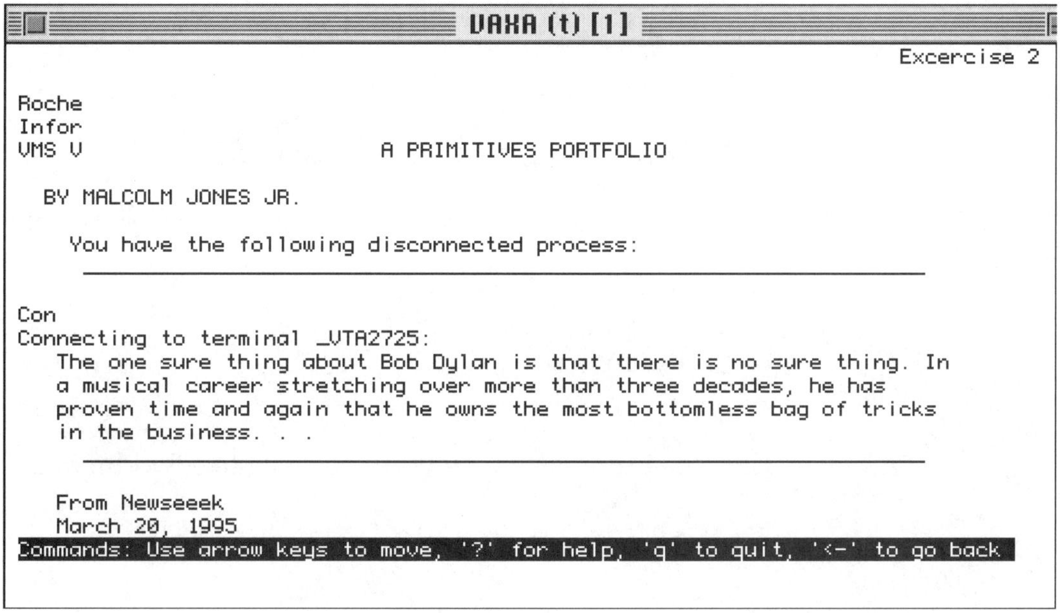

The output in Lynx

Character emphasis

HTML allows you to specify special character highlighting or emphasis (boldface, italics, etc). These elements simply change the typographic rendering of the characters enclosed inside the tags.

HTML allows you to specify these modes in two ways: Logically (by the logical meaning of the special text you wish to mark) and physically (by explicitly specifying the style you want, such as italics, boldface, underline, etc.). Note that the logical styles may not be distinct (i.e. different logical styles may be rendered in the same way).

Logical Styles

The logical styles and examples of their renderings:

EM—Emphasis (usually italics).

```
<EM> this is example text </EM>
```

is rendered:

this is example text

STRONG—Stronger emphasis (usually bold).

```
<STRONG> this is example strong text </STRONG>
```

is rendered:

this is example strong text

CODE—Example of typed code (usually fixed-width font).

```
<CODE> this is example code </CODE>
```

is rendered:

```
this is example code
```

SAMP—A sequence of literal characters.

```
<SAMP> this is example text </SAMP>
```

is rendered:

this is example text

VAR—A variable name.

```
<VAR> this is example text </VAR>
```

is rendered:

this is example text

DFN—The defining instance of a term (often rendered bold)

```
<DFN> this is example text </DFN>
```

is rendered:

this is example text

CITE—A citation (typically rendered in italics).

```
<CITE> this is example text</CITE>
```

is rendered:

this is example text

Thus, <EM>, <STRONG>, <CODE>, <SAMP>, <VAR>, <DFN>, and <CITE> can be used to emphasize or change the typography of designated textual elements in a so-called *logical* manner.

Physical styles

The physical style elements, and their renderings, are:

TT—Fixed width typewriter font.

<TT> this is example text </TT>

is rendered:

`this is example text`

B—Boldface where available.

<B> this is example text </B>

is rendered:

this is example text

I—Italics

<I> this is example text </I>

is rendered:

this is example text

U—Underline (may be rendered as italic in some cases)

<U> this is example text </U>

is rendered:

this is example text

Other formatting markup tags

If a browser does not support these tags, the text will appear normal. Browsers are programmed to ignore (and not display) tags they don't understand. Three of the most commonly supported examples are provided below.

Address

This element is used for address information, signatures, authorship, etc. The rendering of the contents is left up to the browser (indented, italic, right justified).

<ADDRESS> spms@rit.edu </ADDRESS>

is rendered:

spms@rit.edu

Preformatted Text

The PRE element is used to enclose text to be displayed with a fixed width typewriter-like font. This is useful for presenting text that has been formatted for a fixed width character display.

```
<PRE>
Chapter I      The Period
Chapter II     The Mail
Chapter III    The Night Shadows
Chapter IV     The Preparation
Chapter V      The Wine-shop
Chapter VI     The Shoemaker
</PRE>
```

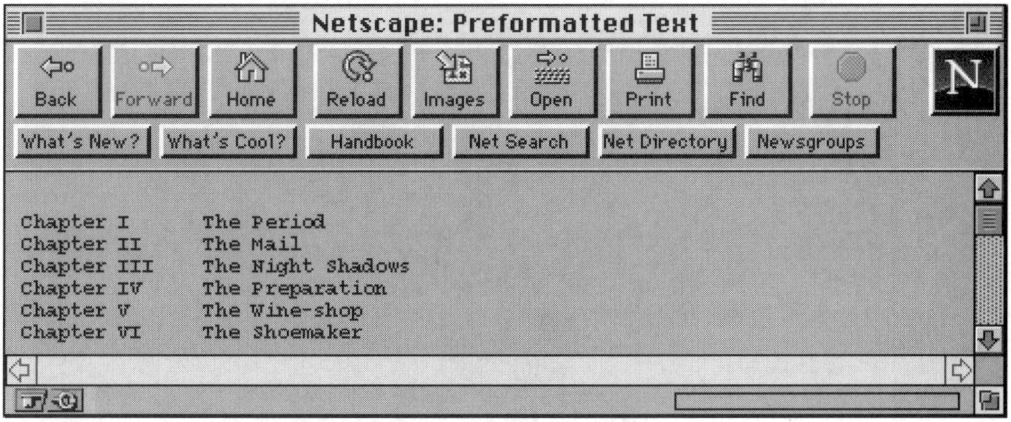

BlockQuote

The BlockQuote element allows quoted text to be rendered in an appropriate way. Typically this may be slightly indented, or italicized. BlockQuote also causes a paragraph break, and typically forces white space both before and after the quotation.

```
<HTML><HEAD>
<TITLE>BLOCKQUOTE</TITLE>
</HEAD> <BODY>
<H1>A Tale of Two Cities</H1><HR>
<I>by Charles Dickens</I><BR>
A story of the French Revolution
<BLOCKQUOTE>
It was the best of times, it was the worst of times, it
was the age of wisdom, it was the age of foolishness, it
was the epoch of belief, it was the epoch of incredulity,
it was the season of Light, it was the season of Darkness,
it was the spring of hope, it was the winter of despair,
we had everything before us, we had nothing before us, we
were all going direct to Heaven, we were all going direct
the other way—in short, the period was so far like the
present period, that some of its noisiest authorities
insisted on its being received, for good or for evil, in
the superlative degree of comparison only.
</BLOCKQUOTE>
```

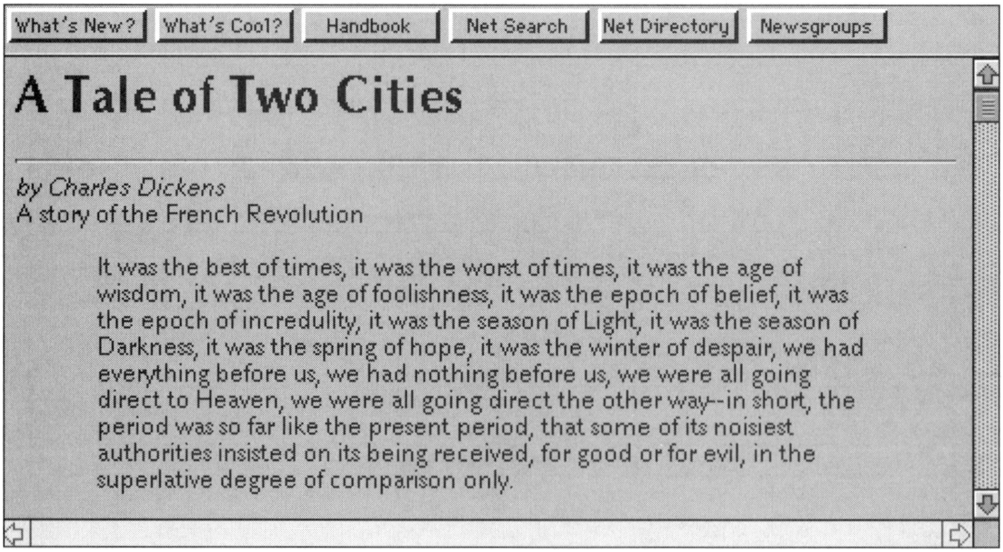

Lists

List tags cause standard items (such as bullets or numbers) to be inserted before the listed text. They can be *nested* within one another to create sub-lists (the effect of this on the appearance of a document differs with the browser that is used). The text of a list entry uses the same style and fonts as normal text.

Unordered List

Using a unordered list tag causes bullets (or the equivalent, if the viewer is using a text mode browser such as Lynx) to be placed before the listed text. An *unordered list* is opened by <UL> and closed by </UL>. Listed items between them are preceded by <LI>.

```
<UL>
<LH>This is a list header!
<LI>Any
<LI>Text
<LI>Here
</UL>
```

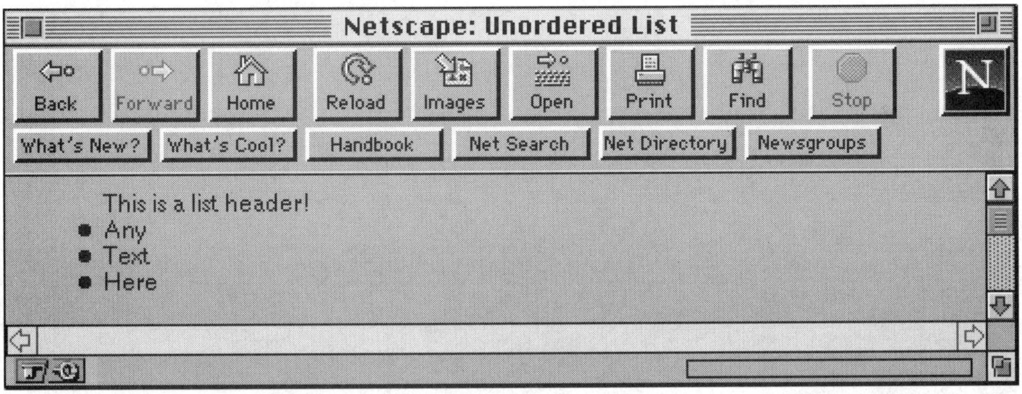

Numerically Ordered List

Using a numerically ordered list tag causes numbers (starting from one) to be placed before the listed text. An ordered list is opened by <OL> and closed by </OL>. Listed items between them are preceded by <LI>.

```
<OL>
<LH>This is a list header!
<LI>One
<LI>Two
<LI>Three
</OL>
```

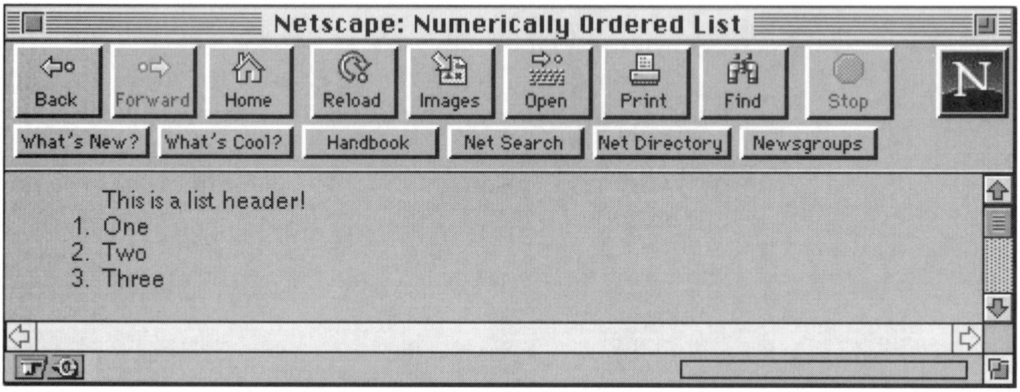

Directory List

This is used for short lists of items, such as file names.

```
<DIR>
<LI>alpha.html
<LI>beta.html
<LI>theta.html
<LI>zeta.html
</DIR>
```

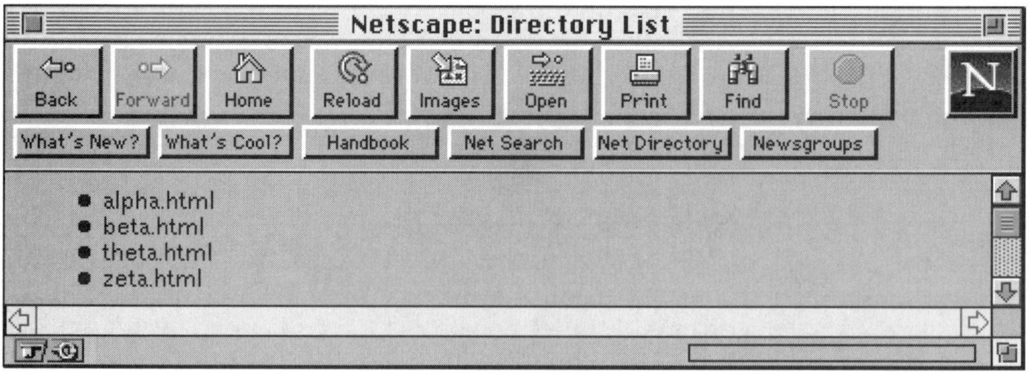

Definition List

Definition lists place nothing before entries preceded by <DT>, and indent entries following <DD>. A definition list is opened by <DL> and closed by </DL>. Listed items between them are preceded by <DT> (Definition Term) and then <DD> (Definition Definition). Use of <DT> without <DD> or vice versa is non-standard.

```
<DL>
<LH>This is a list header!
<DT>Term
<DD>Definition
<DT>Another Term
<DD>Another Definition
<DT>One More Term
<DD>One More Definition
</DL>
```

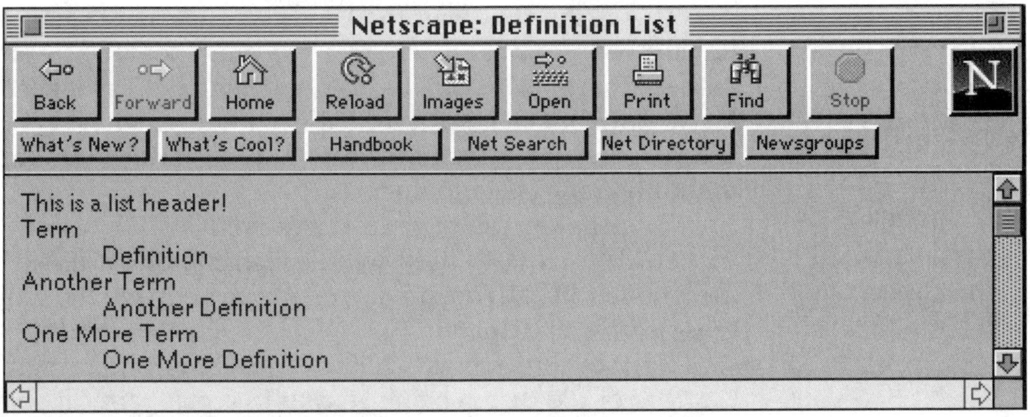

Hot Link

A hot link is a word, phrase or image which is highlighted by your browser indicating to the reader that more information on the highlighted item is available by clicking on that item. When you select or click on a hotlink, you are automatically transferred to the document (or image, etc.) which that hotlink points to (or if the selected resource is not something your browser knows inherently how to display, like a video for instance, an external viewer is launched which knows how to display the selected resource).

Hotlinks are created by placing a word, phrase, or image within a pair of anchors which associate the text with a **URL** (Uniform Resource Locator). A URL is an Internet address. Image hotlinks are called clickable images. A complex image can even contain multiple hotlinks (depending upon where on the image the user clicks), using image maps.

Hot Link concepts

A hotlink tells the browser what type of server is being used to present the file, the address of the machine the file is located on, and where it is on that machine. It also includes a label with which to select the hotlink.

Links to other documents and files appear as highlighted text in a unique style when a page is viewed. When the person using the browser selects the link, the browser is automatically transferred to the file the link points to. The file can be located anywhere in the world.

This is the power of the World Wide Web: transparent linking of information resources to one another across the planet. The person viewing a document does not need to know how or where a resource is stored or what server is used to present it, they just select a hotlink and the file is automatically transferred to their machine.

Three things can happen at this point:

1. The browser itself will display the file (if it is a text file, another HTML document, or a type of image the browser is capable of displaying).

2. If the browser is configured properly, and recognizes the type of file that is being sent to it, a program capable of viewing the file will be launched and the incoming file will be automatically loaded into it (for example, a quick time video or audio clip).

3. If the browser does not recognize what type of file is being sent to it, or it has not been told what type of program to use to view the incoming file, the user will be told this, and asked whether or not they want to download the file.

Hot Link in practice

When creating links in HTML documents the link tag <A . . ./A> is used. The most common kind of linking among

HTML documents is linking to other documents on the Web. Unlike most other tags the <A> tag includes some extra information about the link itself, called attributes. To link to other documents on the Web, the HREF attribute of the Link tag is used. HREF is short for hypertext reference and is used to specify the URL of the file the link points to.

Absolute vs Relative URLs

A relative URL assumes the same access-method, server-name, and directory-path as the document the URL appears in. It indicates the relative position of the target URL from the current URL. For example, here's a URL to a file in the same directory as the document the URL appears in. Below is what a typical <A> tag looks like.

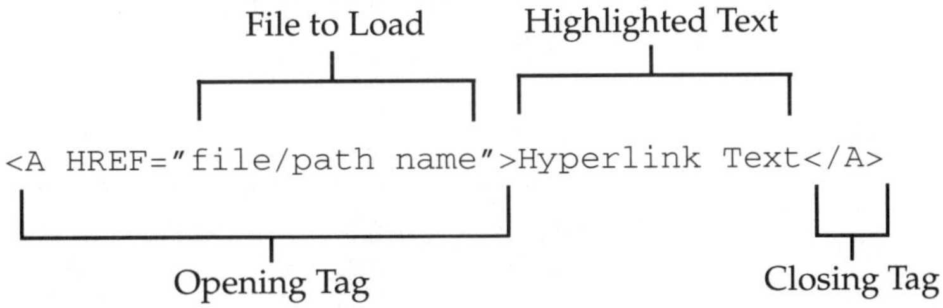

Relative URLs are also sometimes called partial URLs. They are used to point to information resources in the same directory or on the same server. Absolute URLs, on the other hand, are usually used to point to information resources on other servers. In practical terms, this means that you can use relative URLs to direct navigation between documents that you author and use absolute URLs to direct navigation to resources elsewhere on the Internet. A typical link with an absolute path looks like the one below:

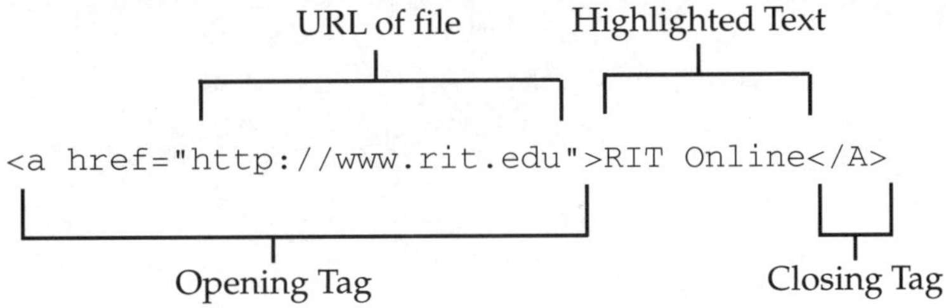

Creating Links

This section demonstrates how to include both an absolute and relative link within an HTML document. Remember that both files need to be in the same directory (folder) for the relative path to work properly. For the absolute path, link to any Web site you wish.

```
<HTML><HEAD>
<TITLE>Linking</TITLE>
</HEAD>
<BODY>
<H1>Linking</H1>
<HR>
<H3>Relative and Absolute paths</H3>
This Relative path connects me to <A
HREF="Exercise 2.html">Exercise 2</A>
<BR>
<BR>
This Absolute path connects me to <A
HREF="http://www.rit.edu">RIT Online</A>
</BODY>
</HTML>
```

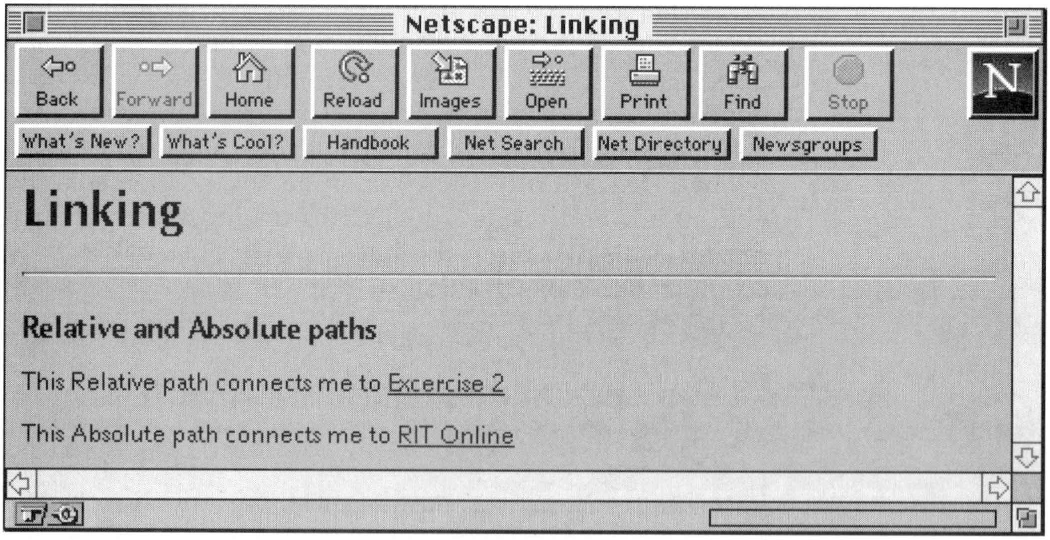

The distinction between absolute and relative URLs is totally hidden from the Web server. When a user selects a relative hyperlink, the Web browser uses the current URL to determine the access-method, server-name, and directory-path, and sends only absolute URLs to the server. Relative URLs also give you the flexibility to move an HTML directory structure anywhere on the Web. And if you move your documents to another server, all the linking remains intact.

Graphics

Fundamentally, there are two different ways to present graphics—inline images and external images. Inline images are displayed by the Web browser as part of your document and are automatically retrieved along with the document. External images are displayed by a separate viewer program (started by the Web browser when needed) and must be specifically requested by triggering a hyperlink.

Inline graphics involve the transfer of a lot of data and so retrieving them can be slow. Many Web browsers let users optionally delay downloading of the inline images. With delayed downloading, inline images must be triggered by a hyperlink before they will be downloaded. When triggered, they still appear inline, not in an external viewer.

When contemplating the use of graphics within your HTML documents, you should consider the user community that will be accessing your Web server. Will they be using graphics-capable Web browsers, such as Netscape, or will they be using line-oriented Web browsers, such as Lynx? The decision is easier if one class of Web browser dominates, but is more complicated if you expect people with both types of browsers to be using your Web server.

Inline images

The inline image is just one more piece of information that is included in the autoflow and autowrap of your HTML on the Web browser screen. In other words, an image is treated just like a word. So an image can appear in the middle of a paragraph. If you want the inline image to stand alone you need to make sure you place <P> or
 HTML tags around it in your HTML document.

GIF

Although many different graphics file formats exist, most browsers will only recognize a few. In practice the only format common to all browsers is GIF (Graphic Interchange Format) devised by CompuServe. While this is the most commonly used format on the Web, other graphics file formats have their strengths and some look set to increase in popularity. Browsers that can display images in other formats, primarily JPEG (Joint Photographers Expert Group) are starting to appear. As yet such browsers are not in common use, but are likely to be so within the next year.

GIF was developed to be a device-independent method of storing pictures. GIF allows high-quality, high resolution graphics to be displayed on a variety of graphics hardware and is intended as an exchange and display mechanism for graphic images. GIF is reasonably well matched to inexpensive computer displays, since it can only store 8 bits/pixel (256 or fewer colors) and most PCs cannot display more than 256 distinct colors at once. GIF does well on images with only a few distinct colors, such as line drawings and simple cartoons. A GIF picture file has an extension .GIF.

JPEG

The JPEG standard is an excellent standard for most realistic images (photos for example, but not line drawings or logos). It uses a powerful, though nominally "lossy", compression method. JPEG is best suited for truecolor original images.

JPEG stores full color information: 24 bits/pixel (16 million colors). Therefore, with full-color hardware, JPEG images look much better than GIFs on such hardware. JPEG files are much smaller than GIFs, therefore, they are superior to GIF in terms of disk space saving and transmission time. A JPEG picture file has an extension .JPG.

Using JPEG for a photographic image for example can produce 10:1 savings compared to GIF, as well as permitting much better display quality on truecolor-capable displays. Netscape handles inline JPEG; most older browsers need to use an external JPEG viewer.

Placing inline images in documents

Inline images are specified in HTML using the <IMG> tag. Like the <P> and <BR> tags, the <IMG> tag has no closing tag. It does however, have flexibility in that it can include up to three attributes; SRC, ALIGN and ALT.

The SRC attribute indicates the filename of the image to be included within the document. It is followed by an equals sign, with the filename in quotes. Be advised that filenames are case sensitive, so be sure to name the file exactly. Also, the image file should be in the same directory as the HTML file. The <IMG> tag looks like this: <IMG SRC="filename.GIF">

```
<HTML><HEAD>
<TITLE>Images</TITLE>
</HEAD>
<BODY>
<IMG SRC="WBradley.GIF">
</BODY>
</HTML>
```

When inline images are autoflowing as part of a paragraph, you can explicitly control the alignment of the image with the text line by using the optional ALIGN attribute of the <IMG> tag. The three values for ALIGN are:

```
<IMG ALIGN=TOP SRC="filename.GIF">
<IMG ALIGN=MIDDLE SRC="filename.GIF">
<IMG ALIGN=BOTTOM SRC="filename.GIF">
```

TOP alignment places the top of the image even with the top of the current line of text, and so on. If ALIGN is omitted, bottom alignment is the default.

```
<HTML><HEAD>
<TITLE>Image/Text Alignment</TITLE>
</HEAD>
<BODY>
<IMG SRC="Horse.GIF" ALIGN=TOP>Text at top of an image
<HR>
<IMG SRC="hglass.GIF" ALIGN=MIDDLE>Text in the middle of an image
<HR>
<IMG SRC="doggy.GIF" ALIGN=BOTTOM>Text at bottom of an image
</BODY>
</HTML>
```

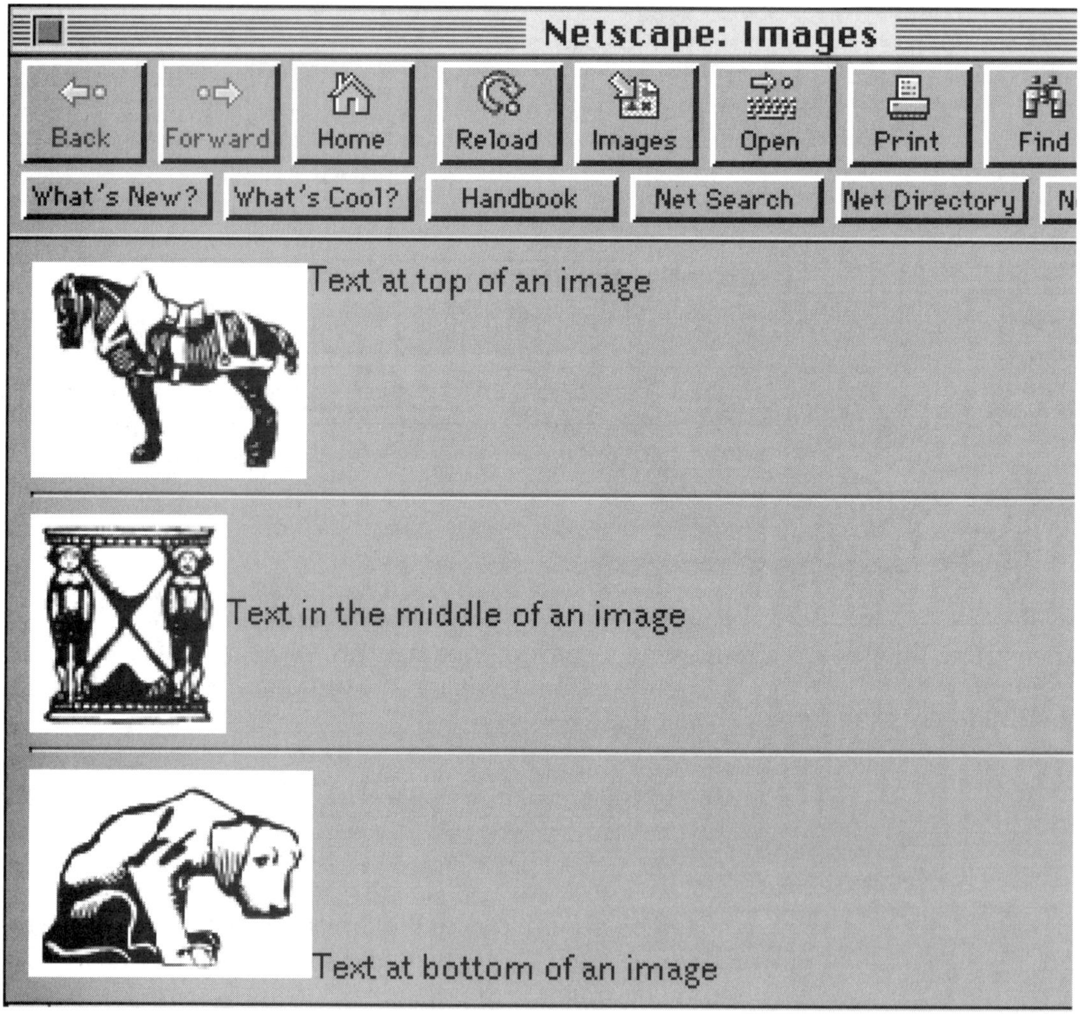

Image Alternatives

Inline images can't be displayed on character-based terminals. Character-based Web browsers indicate that an image is in the autoflow stream by displaying:

[IMAGE]

However, you can override this default and make your inline images more meaningful in character-based Web browsers by using the ALT attribute:

```
<IMG SRC="filename.GIF" ALT="image description">
```

The ALT option makes character-based Web browsers display the image description instead of [IMAGE]. You can also make the character-based browser ignore an image using a null ALT option:

```
<IMG SRC="filename.GIF" ALT="">
```

When a character-based Web browser sees the null ALT option, it ignores the image-insertion tag. The HTML authoring convention is that if you are going to insert inline images in your document, you should at least describe the image with a word or phrase for users with character-based Web browsers.

```
<HTML><HEAD>
<TITLE>Image Alternative</TITLE>
</HEAD>
<BODY>
<IMG SRC="prev.GIF" ALT="[Previous Page]">
<IMG SRC="next.GIF" ALT="[Next Page]>
<IMG SRC="help.GIF" ALT="[Help]>
<IMG SRC="quit.GIF" ALT="[Quit]>
</BODY></HTML>
```

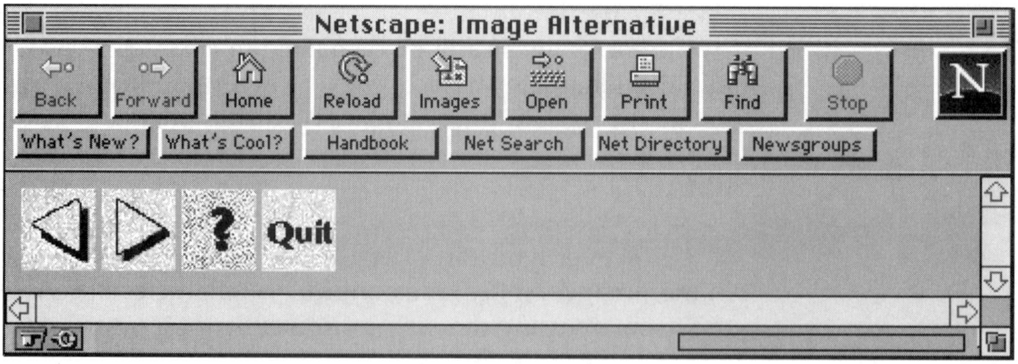

The output in Netscape

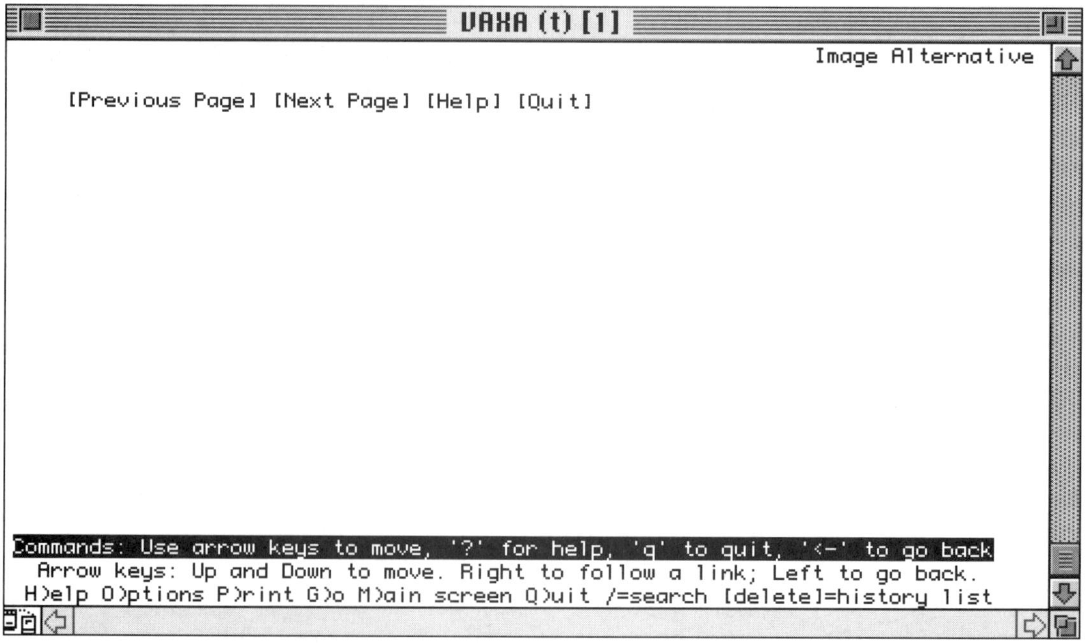

The output in Lynx

Performance

It takes a lot longer for a Web browser to retrieve an HTML document that has inline images than to retrieve one that does not (unless the user specifies delayed downloading). The larger the inline image, the longer it takes. In fact, the time is proportional to the square of the dimension (a four-inch-square image takes almost twice as long as a three-inch-square one). Many graphically-based Web browsers cache inline images. So if the inline image has been used in previous HTML documents in the user's navigation chain, that image might still be cached. Caching also happens when the same inline image appears multiple times in the same HTML document. It is retrieved only once.

External images are images that are not displayed inline as part of your document, but in a separate window, by an external viewer program. This is the technique to use if you have TIFF, JPEG, RGB, or HDF images and don't want to convert them to GIF. This technique is also useful for displaying very large GIF images. Instead of using the <IMG> HTML tag described earlier, include the URL of your external image file as part of the hyperlink. To display a JPEG image in an external window, you can include the following:

```
<A HREF="filename.GIF">hyperlink text</A>
```

The user needs an external viewer installed that knows how to deal with the incoming external image file. Also the Web browser needs to know how to recognize the type of data that's arriving and how to start the appropriate viewer.

Using external images with your HTML document further limits its usability. Just as there are users who can't display inline graphics, there are more who have not configured their environment to support the display of external images.

Images as Links

If you include an <IMG> tag inside of a link tag (<A>), the image specified becomes a clickable link. One trick that's used to solve the performance problem of big images is to use *thumbnails.* A thumbnail is a small version of a figure displayed inline, which is a link to the full-sized image displayed externally.

```
<HTML><HEAD>
<TITLE>Images as Links</TITLE>
</HEAD>
<BODY>
```

```
<AHREF="hi_res.GIF"><IMG  SRC="thumb.GIF"></A>
Click on image to see high-res representation!
</BODY>
</HTML>
```

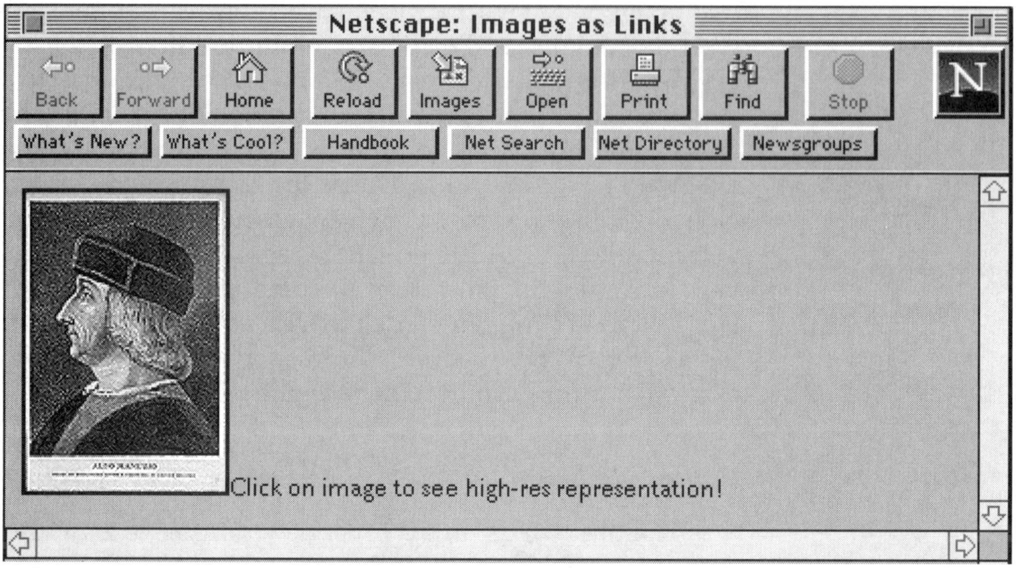

External Media (Video & Sound)

To specify external media in HTML, you create a link to it just as you would another HTML document. In other words, you use an <A> tag with the HREF attribute.

```
<A HREF="external_file.ext">media file</A>
```

The file format of the external media is a **critical** issue when implementing them within an HTML document. The file format signals the Web browser about what type of file is about to be downloaded. The browser then automatically launches an external program that plays the media file.

Formats

The best format for providing sound on the Web is the AU format. The reasons for this choice are: this format can be supported by most of the sound capable machines on the Internet; this format produces a fairly small file that requires little bandwidth to transfer. However, AU files are of only barely acceptable quality, as the 8-bit sample size causes them to sound comparable to transmission over a telephone. Audio files in the AU format have an .AU extension.

Like other files, video files can be identified by their file extensions. There are only few movie file formats that can be viewed from the Internet, which are international standard file formats for multimedia. They are MPEG and QuickTime. MPEG movie formats have the extension .MPG. QuickTime movies have file extension .QT and .MOV. .MOV files can be played on both Macintosh and PCs, while .QT extensions can only be viewed on a Macintosh.

	Format	*Extension*
Audio	AU	.AU
Video	MPEG	.MPG
	QuickTime	.QT or .MOV

Size and Format Indicators

It is considered good etiquette to label all large files (larger than about 100K). This applies to graphics, audio, video, and other large file formats. Because some clients can only accept certain data formats, it also is helpful to provide descriptors or icons that identify a file as image, audio, or video and to indicate its data format and size. This is especially true of very large files. It is extremely frustrating for users to download a 5MB video file only to discover that it is in a format not supported by their browsers. Images as links, a video file, and an audio file.

```
<HTML><HEAD>
<TITLE>Exercise 4</TITLE>
</HEAD> <BODY>
<H1>Exercise 4</H1>
<HR>
<A HREF="sound.AU"><IMG SRC = "sound.GIF" ALIGN=MIDDLE ALT="
">Sample Sound </A>[64k] <I><B>AU Format</B></i> <HR>
<A HREF="movie.mov"><IMG SRC="movie.GIF" ALIGN=MIDDLE ALT="
">Sample Movie</A> [4.3MB] <I><B>QuickTime Format</B></i>
</BODY>
</HTML>
```

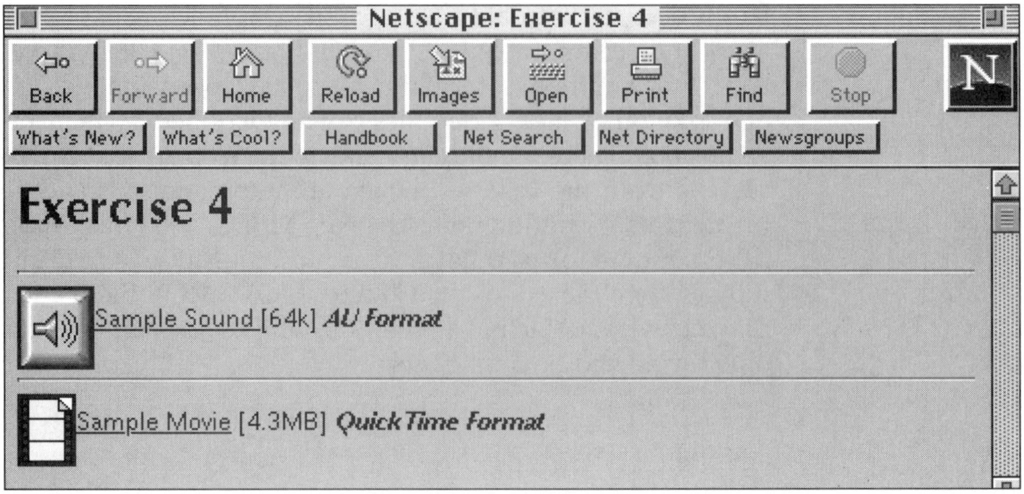

Data size

A 10-second low-quality audio clip can take more than a minute to download over a 14.4 Kbps line. The size of higher-quality audio clips—in stereo or at higher-sampling rates—is at least double that of low-quality clips.

Depending on their quality and duration, video clips generally range from hundreds of kilobytes to 50 Mb or more. Even a few seconds of low-quality video—without audio—can cause significant waiting on a 56 Kbps line. For video as small as 30 seconds, it may take minutes even over a full T1 (1.54 Mb/sec.) connection. Audio and video files should be used sparingly and should be marked as large files.

The best means of reducing the data size for audio or video files is to reduce their quality. Changing from stereo to mono sound can yield a data-size reduction as large as 50%. Lowering the frame rate or resolution in a video file or the sampling frequency in an audio file can substantially reduce the file data size.

Similarly, it is important to use reasonable-length video and audio clips (less than one minute). On some systems, it is difficult to stop a video or audio file from playing once it is started. Therefore, in addition to keeping the data size of the file small, use shorter clips to give your users greater control over what they have to view.

7.

The Internet & the Web

It is made up of more than 15,000 connected networks, has more than 2 million connected computers and has users numbering over 10 million. It is the Internet. It rates special pages in *Time* and *Newsweek* and generates lots of interest but very little business . . . at least for now.

Started in 1969, the ARPANET was the first distributed packet-switched computer network. In 1977, it became one of the Internet's backbone networks, and the protocol research done on the ARPANET was very influential in the development of the TCP/IP protocols currently in use on the Internet. ARPANET's technology became obsolete, however, and it was retired in 1990.

The Internet, which is the world's largest computer network, has been doubling in size (number of hosts and networks) every year since 1988. Once the exclusive domain of research and education groups, the Internet is gaining stature with business users. The Internet is owned by approximately 20,000 organizations worldwide, from large corporations to military services and government agencies. The U.S. government was very influential in the development of the Internet but currently owns or funds only a small fraction.

Most of the Internet is privately owned and consists of local-area networks inside companies. In fact, commercial connections are growing faster than educational ones. Most wide-area Internet connectivity providers are privately owned and operated and will carry any traffic users pay for. Some parts of the Internet (including the fastest parts) restrict use to research or education. Because the Internet had its roots in the open and information sharing world of academia, security has typically taken a backseat. This is an area of concern for business users sending mail over the network.

Password protection is the common security measure on the network today. You can also set up other defenses, including prohibiting incoming network connections, limiting incoming connections to specific services or prohibiting incoming connections for certain hosts or networks for each service. Do not share your password.

With the low cost of connection—often a flat monthly fee for leased line or dial-up access—business users have access to commercial and noncommercial services in the U.S. and 50 other countries. Electronic mail is the most popular application on the network. While on some networks an e-mail message may take hours or days to reach its destination, on the Internet it usually takes seconds to minutes. Internet mail protocols handle queue congestion and flow control automatically.

An Internet message can encapsulate fax, sound, video and PostScript files as well as character sets for foreign languages such as French or Japanese. The addressee needs no special software to receive the message, although some parts of the message can't be interpreted without special software. Through Internet's e-mail, companies can exchange information with other organizations, salespeople, customers and so on with the immediacy needed to conduct day-to-day business. And they can do so on a global basis, often at a cheaper cost than most other e-mail offerings.

Costs came down because Internet services are carried over the Transmission Control Protocol/Internet Protocol (TCP/IP), which multiplexes mail messages (and other services) across common links. This maximizes bandwidth use, minimizing cost and permitting continuous connection. Business users are not confined to one-to-one communications, however; group information exchange happens through the Internet's large base of mailing lists and newsgroups. Mailing lists enable a user to send a single mail message to a mail alias. Software then automatically sends the message to everyone on the list, saving time in retyping addresses.

The Internet is the biggest carrier of newsgroups. Members of newsgroups receive messages and news articles according to their specific group profile. With mailing lists and newsgroups, not only do businesses have the option of getting information such as problem or benefit reports directly from individual customers, but firms can also sponsor customer mailing lists or newsgroups.

Businesses may find it useful to join mailing lists to stay on top of issues affecting their industries. There are lists associated with topics such as technology transfer, foreign business studies and the oil and gas industry, which includes a daily posting of oil prices financial companies may need to make investment decisions. The newsgroup setup tends to be more efficient than mailing lists and can support more participants. The reason is that while a copy of a mail message is sent to each list subscriber, newsgroup messages are sent to each machine that subscribes to the newsgroup. A large number of people in a newsgroup can read the same copy of the message or news article at the same time.

There are newsgroups specific to computer operating systems, other groups that distribute software and still others that discuss specific products. Help wanted and position open newsgroups are among the most popular. For a fee, services enable users to pick and choose among traditional wire service news, sports, features, syndicated columns, business news, newsletters and other packages.

Because the Internet is attached to the global matrix of networks that includes CompuServe, America Online, MCI Mail and Bitnet, among others, businesses sending mail have the potential to reach 30 million people. Internet mail, combined with fast interactive protocols such as TELNET and File Transfer Protocol (FTP), encourage collaborative work over the Internet. FTP and TELNET permit resource sharing by making resources on one machine available across the Internet to users of other machines.

FTP enables users to connect to other computers and perform certain actions, such as listing the files in a directory and copying files back and forth. The TELNET protocol connects users to a remote machine and lets them log on as if they were on a directly connected terminal line.

The use of the Internet to collaborate on producing a book, business document or any other related report saves time, eliminates express mail expense, does away with multiple error-prone manual transcription steps and reduces the need to pass floppy disks back and forth. Vendors are also getting in on the act, cutting down on distribution costs by placing software on the Internet. Apple has made its System 7 operating system as well as early versions of its MacOS available for free by anonymous FTP from apple.com.

The Internet is not about free software or mail functions or supercomputers or any individual service it offers. The bottom line is that the Internet supports the largest and most directly connected community in the world. So many resources are available on the Internet that index and search services are available to help users:

Chronology of the Internet

1969: Bolt, Beranek and Newman (BBN) create a network for the Defense Department's Advanced Research Projects Agency (ARPA) that connects university, military and defense contractors.

1973: The Defense Advanced Research Projects Agency (DARPA) initiates a project on inter-networking to link packet switching networks. There are 35 nodes at this time.

1979: ARPA establishes the Internet Configuration Control Board.

1983: These joined networks become the Internet.

1985: The Internet consists of 100 linked networks.

1989: The Internet consists of 500 linked networks.

1990: The Internet consists of over 2,200 linked networks.

1991: The Internet consists of 4,000 linked networks.

1993: The World Wide Web, created by Tim Berners-Lee, a British citizen working at the CERN Particle Physics Laboratory in Switzerland, goes public with 1,313,000 hosts (systems or nodes, not networks). The Internet is then carrying 5.5 terabytes per month.

1995: The World Wide Web alone carries 4.8 terabytes of the Internet's load. The Internet is turned over to the private sector, no longer applicable for national security purposes.

1996: Over 7 million hosts have access to the Internet, which carries over 25 terabytes of information per month.

Archie

A service that indexes thousands of anonymous FTP servers worldwide—about 150 gigabytes of information. Basically, you ask Archie to locate a program or other package by name, and it responds with a list of all the hosts that have it, which directory and file it is in on each host and when it was last updated.

Wide-Area Information Servers (WAIS)

WAIS can index any piece of text and report to you passages that match key words. It lets you search for information in databases located on servers. Users have access to The Bible, current weather forecasts, documents about the Internet and so on.

Gopher

Gopher helps you find the right menu and keep track of the various servers and information sources. Gopher ties all these items together in a worldwide distributed menu system. Archie, certain libraries and many WAIS servers are available through Gopher, as are text files and software packages. You can get an idea of how gopher works by logging on to one of the public gopher sites.

Commercial companies have extensive research needs, whether it's intelligence gathering on competitors, research to help create a new product or finding out the latest on privacy laws. The Internet lets employees check the Library of Congress, search for relevant book titles and even retrieve information directly, all without getting out of their chairs. The Internet's various indexing and search aids can help locate these resources .

At least two projects are in the works for putting books on the Internet for free: the Online Book Initiative (OBI) and Project Gutenberg. Because of copyright concerns, these services will concentrate on books whose copyrights have expired. Also available will be standard reference books and Internet material.

The Web

The World Wide Web started in 1989 in Bern, Switzerland at the European Laboratory for Particle Physics (known by its French acronym as CERN). It was made so the scientists could transfer hypertext documents (written in HTML hypertext markup language) over the Internet and information could be shared. This was because research was being done at many sites and using the old way of transferring text and graphics on the Internet took too long and was cumbersome. So they started work on a text-mode browser and a graphical browser that would blow away File Transfer Protocol (FTP) and Telnet and image viewers that were used before to get this information transferred. CERN had finished the first of these browsers by the end of 1990 and in 1991 was released for use at CERN. In 1992 CERN began publicizing the World Wide Web Project which made many people interested. This caused more servers to be set up and more interfaces to be created for the Web. At the end of 1993 many software corporations were releasing different browsers and now it is the easiest way to access information on the Internet.

There are over 27,000 sites on the Web in many locations around the globe and it is said to double every 53 days compared to the mere 50 sites in 1993. These sites are created with HTML. HTML not only hyperlinks information and sites, it also supports high-resolution graphics, audio, and video. It also allows Web page designers to organize and layout titles, headings and body text. HTML is what a Web browser translates into an easy to use page of information. In a Web page there may be other links pointing to similar information or recommended Web sites. They are called links because of the hyper-linking that occurs and connects the user to another server or computer. Links are highlighted images or text. Browsers also handle old ways of gathering information that were really hard to understand and use. For example services such as Gopher, FTP and Telnet had to be accessed using UNIX commands which are long and confusing. Most Web browsers can also handle e-mail, USENET groups and Wide Area Information Servers (WAIS).

Web browsers can access lots of information using many different ways, but first you must have a connection and communications software. You need Transmission Control Protocol/Internet Protocol (TCP/IP). TCP/IP is the language of the Internet. Browsers have a few options for a connection. You can get connected through your Local Area Network (LAN) or by the use of a modem to an Internet Service Provider (ISP). If you are going through an ISP you need to use a special communications protocol called Serial Line Internet Protocol (SLIP) or a Point to Point Protocol (PPP). If you are going to connect by modem you definitely need at least a 14.4 bps connection or higher otherwise the download times for pages will be ludicrous. Another thing to keep in mind when using various browsers is their security. If you are going to be doing anything over the Web that uses your credit card or other important information you definitely want a browser which supports either Secure Hypertext Transfer Protocol (SHTTP) or Secure Sockets Layer (SSL) standards.

Some Web browsers have more features that others and some work more efficiently than others. As time has gone on improvements are being made to many of the browsers and connection speeds are increasing, making it easier to access information in a shorter period of time. One of the first browsers to come out was Mosaic. It was created by computer science students and the University of Illinois; National Center for Supercomputer Applications (NCSA). It was the first program to take Web files and their links and display them in a graphically appealing manner. When Mosaic was introduced it shifted the use of the Web from scientists and academics to commercial and consumer users. Mosaic started as a freeware program. Its older versions are still free, but they are not working as efficiently as the commercial versions. Spyglass Software released the commercial version as Enhanced Mosaic. This version of Mosaic supports multiple downloads and allows the user to go to another page while it downloads another. Enhanced Mosaic also has security. It supports private key encryption and SHTTP. This version of Mosaic also has a better toolbar and you can stop downloads and reload a current page.

In late 1994 Netscape Communications introduced Netscape Navigator. It was written by most of the people that worked on NCSA Mosaic and was led by Marc Andressen who had graduated from University of Illinois. It is said that more than 75% of the people that access the Web use Netscape. Netscape downloads text before graphics. It can also perform multiple/simultaneous downloads. You can turn off the graphics so downloading a page is quicker. You then have the choice to reload the page with graphics. Netscape also gives the user more security than other browsers. In addition to supporting its own Secure Sockets Layer scheme it also warns you when the data you are transmitting is not secure. When you first launch Netscape Navigator it takes you to the Netscape Communications home page. At this point you can perform Web searches and get updates from Netscape. Netscape also has a feature known as a bookmark. When a user finds a site they like they can mark that page with a bookmark and it is more easily accessible later. Netscape also makes it easy to navigate through USENET groups.

There are also many little extensions for Netscape that will help it handle other file formats during downloads. Netscape is freeware but if you pay for the commercial version you get a manual and tech support. Netscape also recently signed deals with credit card companies that will allow for security on their commercial version only when making financial transactions. Adobe and Netscape have integrated the portable-document format software, Acrobat, into Netscape Navigator. Adobe made a plug-in for Netscape called Adobe's Acrobat WebLink that will allow users to link PDF documents together on the Internet. Future versions of Netscape will be able to display Acrobat PDF documents. Netscape has also worked with Macromind on a program called Shockwave. This program will be integrated into Netscape and allows for cross platform playback of Macromind Director files. Since Netscape does control 75% of the market it recently had amazing things going on when its stock debuted, considering it is only a two-year old company.

Yellow Pages envy

The McKinley Internet Directory is a selective hyperlinked, online directory to the Internet a reference work of cyberspace. Selectivity also aids the *seamless linking* of advertising and content. An icon placed next to each Internet information resource leads directly to businesses that sell related products. More than 1,000 advertisers were lined up in advance of the directory's early spring launch.

Every RBOC is evaluating Internet opportunities for Yellow Pages because it is a phenomenally profitable business. On average, telephone companies earn 20% of their net profit from Yellow Pages advertising. Yellow Pages ad revenue totaled over $11 billion in 1996.

Online advertising

It is the answer to an advertiser's prayer. In a champagne ad from the latest issue of a magazine on the Web, readers click their mouse to play a game, hear a joke, order merchandise, or search for drink recipes. It's enough to keep a reader busy in the brand message for perhaps half an hour. But another button lurking on that PC screen also makes such interactivity an advertiser's nightmare. Labeled "Back" it causes the ad to disappear as quickly as it arrived.

Full of promise, a small revolution in advertising is drawing marketers from Coke to Toyota to the Internet, online services, CD-ROMs and interactive television. Still a fly speck in the $150 billion U.S. advertising industry, interactive promotions have risen from virtually nothing in 1994 to well over $100 million in 1996 and will swell to $4 billion by the year 2000, according to a research firm. Interactive media could rival broadcast TV's current ad revenues of $30 billion during the next decade.

Madison Avenue has awakened to new media. Just two years ago, virtually no major ad agency had a group assigned to interactive advertising. Today, all of them do. Still, they are competing with specialized agencies that have spent as much as a decade designing interactive sales demos, computerized shopping information kiosks and promotional diskettes. To become bigger players in the interactive arena, some established agencies have made acquisitions. Last year, for example, D'Arcy Masius Benton & Bowles, one of the world's largest ad agency, bought a well-known interactive agency, Einstein & Sandom.

Clients are eager to begin interactive advertising. Fifty percent of major U.S. corporations have plans for advertising or promotion on the Internet's World Wide Web—in addition to the 9% already marketing their products there—according to a survey by market researcher Yankee Group. The Web's draw is its environment of full-color graphics and the ease with which users can jump from one site to the next by simply clicking on a highlighted word or an image. To lure customers to their sites, companies buy "billboard" or "Web pointer" ads on more-popular Internet locations.

Auto makers were among the first to embrace interactive advertising. The Internet now has sites promoting 54 different makes of vehicle—from Accura to Winnebago. The *Popular Mechanics New Car Buyers Guide* on CD-ROM has interactive ads from Cadillac and Toyota. And Detroit's Big Three have all marketed their products within the major online services. (*U.S. News Online*, the first media area within CompuServe to carry advertising, has ads from Saab and Subaru. The magazine's Web site also includes ads.) Cars fit into the interactive environment because people like to research before making such a major purchase.

The challenge of interactive advertising is to attract and hold the attention of consumers who may care little about the advertised brand—and who don't have to sit through a commercial just to see the rest of a TV show. The solution, say most advertisers, is information AND entertainment. MCI, for example, sponsors an interactive soap opera on the World Wide Web called "Gramercy Press." Readers follow the office drama by snooping on the characters' voice mail, e-mail and other office communications, all in a subtle promotion of these MCI services.

Interactive advertisers are not content to wait for customers to come to them. A growing number of companies are directly approaching consumers with highly targeted online marketing efforts. The Internet abounds with sweepstakes, questionnaires and areas of special access, which require users to give their names, e-mail addresses and other data.

There are other concerns, too. The Internet and commercial online services reach homes and businesses over phone lines that lack the capacity to carry video. CD-ROMs can't be updated and don't allow customers to talk back directly to advertisers. And interactive TV is stuck in costly and long-delayed trials. To fully blossom, it will need high-capacity, video-capable networks connected to home PCs and TVs. The wiring and equipment for interactive TV could cost $1,000 per home, but surveys find that TV watchers are not willing to pay much more than their current cable bill for such a service. As they already do in magazines, newspapers and TV, advertisers will have to cover the difference. Video-on-demand movies, for example, may cost $4 without ads and $1 for those willing to sit through advertising. The interactive era will offer consumers the opportunity to tune out ads, but consumers may ultimately find that they would rather tune in.

Newspaper and magazine publishers have been the first to stake claims on the Internet. They are still tinkering with the advertising models for their fledgling World Wide Web sites and many publishers ventured into cyberspace assuming that advertising alone could carry the load. Advertisers have been skittish about pumping money into a medium with an audience whose size and habits are nearly impossible to gauge. Although several publishers with Web sites, from *New York Times* to *USA Today*, can say they're ahead of advertising projections, none have turned a profit.

Publishers are developing new economic models that supplement ad revenue by making users pay. After a year and a half on the Web, Time Warner's Pathfinder site is offering its first subscription product. *The Wall Street Journal* announced its Interactive Edition and plans to start charging consumers for the service.

Getting users to pay up won't be easy. "It's the Net culture: They like free stuff," said the president of *New York Times Electronic Media Co.* "Once people get something for free, it is difficult to get them to pay for it."

In recent years, publishers have expected ad revenue to cover their Internet costs and have resisted charging users for fear of limiting their audience. But as publishers become more adept at working in cyberspace, they are setting up premium zones on their sites that charge an admission fee. So far, only a few prominent sites, such as *ESPNet SportsZone*, have been successful in charging users for access to their most valued news or information. Publishers are hedging their bets. *The New York Times* will charge users for different services on its site, while the *Boston Globe*, a sister publication, has no plans to charge users of its Web site.A hybrid of free material supported by ads and premium material with an admission fee is quickly becoming popular with publishers on the Web. Advertising with the right mix of niche products could bring a profit.

Gannett scrapped its *USA Today* electronic version on CompuServe, which levied a subscription fee after lagging sign-ups; it peaked at 1,600 subscribers. Now it offers a free site on the Web supported by ads and soon plans to offer the newspaper archives for $9.95 a month for 100 articles. Time Warner's Pathfinder will include Pathfinder Personal Edition, which will let subscribers pick the news they want from *Time* titles. Knight-Ridder's Mercury Center, created by the *San Jose Mercury News*, dangles headlines and news summaries free, but charges nonsubscribers of the print version $4.95 a month for full stories and certain columns.

Advertising on the Internet is expected to be $44 million in 1996 and could reach more than $2 billion by the year 2000; advertisers long have been skittish about committing money to on-line services, largely because advertisers aren't sure who's reading them. Newspapers and magazines are falling over themselves to publish on the Internet, but doubts remain whether potential readers will find computer delivery attractive compared with traditional methods.

Subscriptions to newspapers and magazines in Europe, delivered across the world-wide network of personal computers, will increase by 150% a year between now and 2000, according to the technology consultancy Datamonitor. German media giant Bertelsmann said its digital media sales would hit $1.38 billion by the year 2000, boosted by its linkup with online service America Online. Even software giant Microsoft Corp. is offering its customers news and multimedia services. But some analysts wonder whether people are willing to pay significant money to get news on a computer. They say newspapers and magazines deliver a sophisticated, user-friendly and low priced product perfected over decades. Many pioneer Internet users seeking a news briefing have returned to traditional methods saying connections are maddeningly slow and systems often impenetrable. Dow Jones & Co. Inc., publisher of the *Wall Street Journal*, announced the launch of its Interactive Edition. This will be delivered across the Internet and promises updated news 24 hours a day, stock prices, and background information—it is aimed at small and professional investors. This was a free Internet service but the new Interactive Edition will cost $49.95 a year. The point is to get a whole new class of people—those who are searching for information, small investors, professional investors. Bloomberg and Reuters are already in the market. About 2.6 million households will have signed up for online news services in Europe by the year 2000. Like many projections in these new computer driven markets, they show extraordinary extrapolations from current sales.

Hundreds of newspapers across the globe have published on the World Wide Web. But many just dump the print edition on to the Net without tailoring it for electronic publishing. This is called "shovelware" and explains why most newspapers are available free on the Internet. Persuading people to pay for information on the Internet may be the real challenge.

Getting connected

To get connected, you'll need to sign up with an Internet service provider. The company you select provides your on-ramp to the Information Superhighway. Use a local provider to avoid being charged for long-distance telephone calls in addition to usage fees or monthly charges. Contact the providers listed under your area code and find out what types of Internet access they offer, and at what cost.

There are four ways to connect to the Internet—two indirect methods and two direct methods. Your choice should be determined by the features you need.

Indirect connections

Consider an indirect connection if you're connecting a single computer and need access only to the subset of Internet features that are available through the software you use, or need access to all the features at the lowest possible price. Either way, you'll be making a trade-off between price and ease of use. If you don't use an online service, you'll need to learn the basics of the UNIX operating system.

Mail-only connection

The most basic option for connecting a single desktop computer—mail-only connections—are offered by many providers and by nearly all the major online services, including AppleLink, America Online, CompuServe, GEnie, and Prodigy. Your local bulletin board service (BBS) may exchange e-mail with the Internet; all Fidonet and many FirstClass BBSs let you send mail to Internet users, and often offer this service free.

UNIX shell account

This type of service involves accessing the Internet through a command-line user interface, and is often used to connect a single computer. The interface looks and feels like UNIX—usually because it is. With a shell account, you're not directly connected to the Internet, but are sharing a UNIX computer that is. You use your computer as a terminal to operate the character-based client programs that provide access to Internet services; the programs run on a remote UNIX computer.

Since you are using a remote computer, a task such as downloading files is a two-step process: The first step is to get files from the Internet to the UNIX server, and the second is to get them from the server to your computer.

Direct connections

With a direct connection, either your computer or your local area network (LAN) is actually "on the Internet." If you're already connected, here's a simple way to find out if you're "on the net": If you can initiate or receive a file transfer using a tool such as Archie, Fetch, or TurboGopher, your computer is on the net, and can be an Internet client, an Internet server, or both.

Dial-up connection

With this type of connection, you can establish a link to the Internet from a single computer, using a modem and one of three dial-up LAN technologies: Serial Line Internet Protocol (SLIP), Point to Point Protocol (PPP), or Apple Remote Access and a DDP/IP gateway. The technology you use will be determined by the technology your Internet service provider uses to answer your call. In any case, you'll be connecting to a LAN that is already on the Internet.

Your service provider may want to know whether you want to be connected using SLIP or PPP—both provide equivalent Internet access. Implementations of SLIP and PPP are extensions that allow MacTCP to be used with a modem. Several free and commercial software packages for SLIP and PPP are available.

PPP lets you use multiple protocols concurrently over a dial-up link, and many organizations are starting to use it as a dial-in solution for Macintosh computers, IBM or IBM-compatible computers, and UNIX workstations. Since SLIP supports only a single protocol (TCP/IP), PPP is your best option—provided you have a choice.

Dial-up software requirements

A dial-up connection gives you a wide choice of Internet applications. You'll need a 28.8 bps modem and an account with a dial-up Internet service provider.

Apple Remote Access and a DDP/IP gateway, is favored by organizations that want to extend their Internet connection to remote or traveling users. Apple Remote Access servers, used with the new Apple IP Gateway software (or a third-party DDP/IP gateway), make it easy to provide Internet access to remote users. Of the dial-up Internet solutions, it's the easiest to configure. On the remote Macintosh, it allows MacTCP to work without any extra software, the same way MacTCP works on LocalTalk networks.

Connecting your LAN to the Internet

This solution lets you connect a group of computers, and is ideal for an organization that wants to publish on the Internet. You don't need a modem for this type of connection. Each PC connects to your site's LAN using MacTCP and LocalTalk, Ethernet, Token-Ring, or whatever cable system is installed at your site. The LAN is connected to your Internet service provider through a router (usually using a high-speed dedicated line); the connection is maintained 24 hours a day.

You can also install Internet server software on computers on your LAN, and share information about your organization with other Internet users. Macintosh computers with software such as FTPd, GopherSurfer, or MacHTTP are the easiest Internet servers to set up and administer. FTPd provides the server side of Internet file sharing. GopherSurfer and MacHTTP allow remote users with Gopher or World Wide Web browsers (such as Netscape) to navigate through the data you've published.

Notes for network managers

If you are connecting your LAN to the Internet, you or your network manager will need to work with your service provider to obtain a group of Internet Protocol (IP) addresses. An IP address is a unique number that allows a network device such as a computer or printer to be recognized on the Internet. Information that identifies your site is composed of two parts: a group of unique IP addresses and a domain name, which is the site name seen by other Internet users. The combination of your site's domain name and IP addresses allows mail and other types of traffic to be directed to a router that can forward messages to your site.

Most organizations want their own domain name, and you will need to find out whether the name you have chosen is already in use. Work with the Internet Network Information Center to get the IP addresses and to verify the availability of your domain name.

The role of the router

For data to be routed to the appropriate destination, your network must be connected to the Internet. The router provides the physical connection between your LAN and a line to your service provider; it directs communications to and from your network. Any router that supports TCP/IP—the protocol used on the Internet—will ensure compatibility. Just remember to check with your service provider before purchasing equipment, to make sure your provider can help you get it configured and running properly.

What can you do on the Internet?

The most basic service is e-mail, and an online service can usually meet this need. More advanced features include everything from videoconferencing with other Internet users to exchanging files directly with another Internet user anywhere on the network. Once you're connected, you're all set to use the Internet to communicate with other users—across the hall or across the globe.

In deciding how to make your connection to the Internet, you will make trade-offs, depending on your needs and your budget. There are many "deals"—most of which provide unlimited connection time for a flat monthly rate.

The Internet is a vast community of interconnected computer networks. It's used by millions of people in 81 countries, and it's growing at a rate of 10% to 15% per month. Once every 5 seconds a new person signs on. The "net," as it's called, is a truly global resource, with prodigious amounts of information on art, anthropology, engineering, finance, literature, marketing, music, and virtually every other area of interest imaginable.

The Internet is used by people in all fields—not just by the academics and government researchers who founded it. The Internet enables teachers to bring the world into their classrooms; students can go online to use the Library of Congress, collaborate with peers halfway around the world, download weather maps and images from NASA space explorations, or exchange e-mail with researchers in Asia. For businesses, the Internet represents the opportunity to let millions of people know about their products and services.

From the smallest school to the largest corporation, anyone who is connected can publish information on the Internet—in effect, making that information available to all 25 million users in 81 countries.

The Net is not a life

Some see the Internet as the embodiment of the future of communication, and if it provides universal access, it will. But more people are writing about the net than actually using it.

Back at the time of Gutenberg, only a privileged few could afford any kind of book. Today Internet access rests with those earning $57,000 year or more. Just as the book and other forms of print became more democratic, so must Internet access. Gutenberg's invention made lead more valuable than gold because virtually everyone understood what print could do. No one really knows what the so-called information superhighway can do.

For print to be displaced as the primary means of communication, you must replace it with as pervasive a communication channel. Cable TV does not do it. Broadcast TV is close. The Internet is not even close. It is more or less electronic graffiti on the info highway.

"Surf the web" says a lot about people with attention deficit disorder. The Internet could be a network of networks with a few million people with short attention spans and too much idle time searching for the ultimate diversion. Most Internet users are white males under 24 and popular sites involve erotica. The reason: no staples.

The Internet is like CB radio, only you have to type. It's nothing to e-mail home about. How do you make money on the Internet? This guy ran magazine ads for years telling you how he got rich and for $20 he would tell you the secret. I sent the 20 bucks and got a booklet telling me to run magazine ads on how I got rich by charging $20 for the same information. More people will make more money telling other people how to make money than real users will, trying to make money by actually doing something. Another oxymoron: net profits. Some say cyberspace is a parallel universe. Get a parallel life.

8.

CD-ROM & DVD

After almost a decade, CD-ROM technology is now a standard component for desktop computers. Just as CD-ROM is becoming a mainstay it is also nearing the end of its useful life as a new generation of devices emerges to replace it. The *ROM* in CD-ROM for read-only memory is ideal for large amounts of information that does not change. Newer CD drives and disks will be written repeatedly like high-capacity diskettes and will bring great possibilities as backup devices as well as digital media distribution products.

Someday there will be no difference between CD-ROM and video/audio CD formats which are largely incompatible today. This upcoming enhanced CD standard will lead to lower prices and software, audio, text and video will be tightly integrated, permitting interactive music video disks. Most CD-ROM drives in use are *double speed* or 2X drives, that spin at twice the rate of the first generation, and thus send data to the computer more quickly, improving video and sound performance. Triple speed or 3X was eclipsed by quad speed or 4X drives, which are now common.

Most computers ship with quad-speed drives and manufacturers have already stopped making double-speed drives. The next leap in storage capacity will allow CD-ROM drives to store 3 to 4 gigabytes of information on a single disk. Even with the best data compression, the current CD-ROM disks can hold only about an hour of video but at 3+ Gb, a feature film could be stored on a single high-density CD-ROM disk. As 3-4 Gb disks become widespread they will replace quad-speed drives with something even faster. The combination of higher-speed drives and higher-capacity disks opens the possibility of laser-quality movies on a single disk, which in turn could kill VHS tapes.

CD-ROM itself could be displaced by advances in broadband networks and alternative storage technologies. Software, including video and audio, will be transmitted over networks, whether the network belongs to the phone company, the cable TV company or the satellite company. The issue is bandwidth, or capacity of the data pipeline, and transmission time. It can take hours to download video on a 14.4 kilobit per second telephone line. Until the rest of the information superhighway catches up, we need storage capacity. That storage may or may not be on CD-ROM. Even in quad-speed drives, a hard disk is 100 times faster than CD-ROM, and RAM is 1,000 times faster. CD-ROM software developers are becoming adept at copying parts of their programs into computer memory. We could have three gigabytes of RAM in personal computers, and gigabyte hard disks are already common. Some sort of hybrid system involving the static storage of data on CD-ROM, linked to the immediacy of an online service could be an intermediate approach.

Compact disk formats

In the beginning, there was CD-DA (Compact Disk-Digital Audio), or standard music CDs. CD-DA begot CD-ROM when developers realized that you could store a whole lot of computer data on a 12 centimeter optical disk (650 Mb). CD-ROM drives are simply another kind of digital storage media for computers, albeit read-only. They are peripherals just like hard disks and floppy drives. (Incidentally, the convention is that when referring to magnetic media, it is spelled "disk." Optical media like CDs, LaserDisk, and all the other formats are usually spelled *disc*.) We use *disk* for everything.

CD-I (Compact Disk-Interactive) came next. This is a consumer electronics format that uses the optical disk in combination with a computer to provide a home entertainment system that delivers music, graphics, text, animation, and video in the living room. Unlike a CD-ROM drive, a CD-I player is a standalone system that requires no external computer. It plugs directly into a TV and stereo system and comes with a remote control to allow the user to interact with software programs sold on disks. It looks and feels much like a CD player except that you get images as well as music out of it and you can actively control what happens. In fact, it *is* a CD-DA player and all of your standard music CDs will play on a CD-I player; there is just no video in that case.

Next came CD-ROM/XA (eXtended Architecture)—a CD-ROM drive but with some of the compressed audio capabilities found in a CD-I player (also called ADPCM). This allows interleaving of audio and other data so that an XA drive can play audio and display pictures (or other objects) simultaneously. There is special hardware in an XA drive controller to handle the audio playback. Still with me?

Then comes the Kodak Photo CD—digital pictures on compact disk. They teamed up with Philips to develop the standard for Photo CD disks. At this point, a new problem enters the picture, if you'll pardon the expression. All of the disk formats mentioned so far are read-only; there is no way for anyone but the producer of one of these disks to store their own content on the disk—that is, to write to it. But there already existed a technology called WORM (Write Once Read Many). This is an optical disk that can be written to just once. You can *burn* data on it, but once burned the data cannot be erased, although it can then be used like a CD-ROM disk and read forever (depending on your definition of how long forever is).

CD-ROM, CD-ROM/XA, and CD-I disks are normally "mastered," as opposed to burned. That means that one master copy is made and then hundreds, or thousands, or millions (if you need that many) of replicates (a fancy word for copy) are pressed from the master. This process is much cheaper than burning for quantities above a few dozen or so. Generally, disk manufacturers can handle all of these formats since the underlying technology is the same; the only difference is in the data and disk format.

The reason that WORM technology was critical for Photo CD is obvious—the content of these disks is not determined by the manufacturer or publisher. For PhotoCD, each disk will be different—a roll or a few rolls of film per disk from a customer.

Kodak and Philips wanted PhotoCD disks to be playable on both computer peripherals for desktop publishing uses *and* on a consumer device for home viewing. For the former,

CD-ROM/XA was chosen as a carrier and for the latter CD-I, which was already designed as a consumer electronics device, and dedicated Photo CD players. This desire for a hybrid disk, or one with multi-platform compatibility, led to the development of the CD-I Bridge disk format. A Bridge disk is one that is readable on both a CD-I player and a CD-ROM/XA drive. Disk makers can handle all of these formats as the underlying technology is the same; the only difference is in the data and disk format. CDeja vu?

This Bridge format is the reason there is so much confusion about CD-ROM drives for PhotoCD. A drive that supports PhotoCD must be a CD-ROM/XA drive that is also Bridge-compatible. (The technical description of Bridge disks calls for supporting certain kinds of sectors identified by "form" and "mode" bits, which is what you usually hear instead of the "Bridge" disk nomenclature.)

Although a WORM disk can only be written to once, it is not necessary to write, or burn, the entire disk all at once. You can burn the disk initially with, say a few hundred megabytes of data, and then go back later and burn some more data onto it. Each burn must be to a virgin part of the disk; once a spot on the disk is burned, it cannot be re-burned. Each burn operation is referred to as a "session," and a drive or disk that supports this multiple burning operation is called "multisession." I think I've got it.

Originally, WORMs were single session only. That is, you could not go back and add data to a WORM disk once it was burned, even if it was not full. For PhotoCD, Kodak wanted the consumer to be able to add more pictures to an existing disk as additional rolls of film were processed. So the extension of WORM technology to multisession was developed and adopted for the Bridge disk format. This required hardware changes to CD-ROM/XA drives and that is why there are a small number of single session XA drives on the market and multisession ones now commonplace.

A single session drive can read a multisession disk, but it can only read the contents of the first session that was burned. Incidentally, all Philips CD-I players are multisession, although all current CD-I disks have only a single session on them.

A massive shakeout is ahead for hundreds of CD-ROM content developers. CD-ROM marketers face steep marketing challenges in differentiating their wares from thousands of new offerings on every imaginable subject from entertainment to education to self-help to erotica. Despite these difficulties, the long-term outlook for CD-ROMs is bright—that is, for marketers who survive the coming shakeout. The number of CD-ROM titles published this year will nearly double to 4,000 over last year's 2,400. Other industry observers are even more bullish, saying the figure may be as high as 7,000.

Projected Worldwide Shipments of Data Drives in Thousands

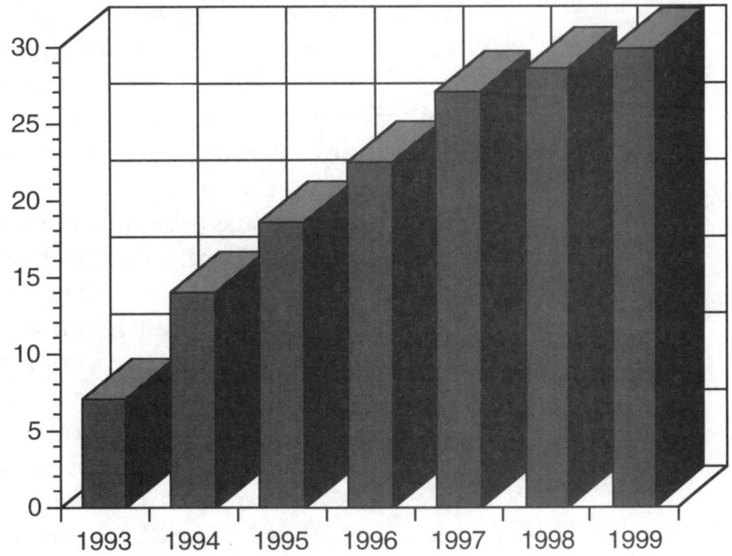

To summarize

The CD-ROM is physically identical to an audio CD. But a format initially designed to deliver an hour of high-quality audio is hardly ideal for conveying a similar quantity of richer video data, complete with sound. And laser disk and CD-ROM couldn't be less alike. Laser disk offers the best picture image now available, while the digitized moving picture on computer is still far from perfected.

Even though it is a decade old, CD-ROM technology has only recently become a standard component in new personal computers. And it is is already past its prime just as it is becoming widespread. New generations of devices are emerging that will gradually, but inevitably, replace it. New standards will eliminate the difference between CD-ROM and audio CD formats. The two are largely incompatible today. One can play audio CDs in a CD-ROM drive, but not CD-ROM disks in an audio CD drive.

Most computers ship with quad-speed drives and the next jump in storage capacity will allow CD-ROMs to store 3.3 gigabytes of information on a single disk. The combination of higher-speed drives and higher-capacity disks opens the possibility of laser-quality movies on a single disk, which in turn could hasten the demise of VHS videotapes. When the rewriteable multi-gigabyte optical disk drives reach the consumer market in 1998, the VCR itself will begin to fade.

The new disk format goes by the name of DVD, or digital video disk. A single disk can hold a 135-minute feature film complete with three separate language tracks of surround-sound audio and four sets of subtitles. Picture quality can visibly surpass what videocassettes deliver. DVD players will also read current compact disks. Eventually, portable players could be no bigger than paperback books.

A DVD-ROM, as the first incarnation will be known, can hold as much as 4.7 gigabytes of information on a single side, about seven times the capacity of current CD-ROMs. Double-layered DVDs will hold 8.5 billion bytes, and double-sided disks will double those numbers. Data can flow steadily from the disks at a rate of 1.35 megabytes per second, more than twice as fast as 4X CD-ROM players. Recordable, and later rerecordable, DVD disks are expected to appear.

Eventually, DVD and CD-ROM could be displaced, or even replaced, by advances in so-called broadband networks and alternative storage technologies. Software, including video and audio, could be transmitted over high-speed networks, whether the network belongs to the phone company, the cable TV company or the satellite company.

9.

Intellectual Property

We are surrounded by text, sound, and graphics that are ripe for the plucking. The World Wide Web Browsers let you download all manner of material. But is it free?

The ability to store, transmit and access text, data, sound, images, and other types of information in digital form has led to the explosion of the multimedia market. With this rise in use of multimedia, the integration of the computer, communications, entertainment, music, and publishing industries has become more apparent.

Since multimedia depends on the contribution of different technologies from a variety of industries, the development of laws to protect the development and distribution of this media have been raised. These new laws will probably strain the already existing computer laws to their limits. This summary will provide an overview to the legal issues that arise with respect to the development, distribution, and protection of the new form of communication known as multimedia.

Defining multimedia

A multimedia product includes text, sound and images that are assembled and then accessed interactively by the user via their computer. The user is prompted to use the mouse or keyboard to "click" from screen to screen.

Multimedia is a combination of software and multiple forms of content, in digital form, stored and delivered using computer technology, and used in an interactive and nonlinear manner. Multimedia is like having a book with lots of pages and no page numbers, that talks to you, and does what you tell it to do.

First of all, this chapter is not offering nor does it claim to offer legal advice. Consult an attorney, especially one who specializes in copyright law. The material presented is derived from a number of sources and attempts to offer an outline of current copyright practices. It is neither complete nor comprehensive. And lastly, do not assume that any of it is right. Only Courts determine that.

Software plus content or content plus software

Digital media contains different combinations of sound (usually spoken words, music, or sound effects), images (such as photographs, graphics, motion picture clips, or animation), and some sort of software to run the program. This software is used to make the program actually operate. The result is a seamless production of content and form.

Stored and delivered via computer technology

Since digital media products are *digital* in nature, they are usually produced in machine-readable form. This can be anything from a magnetic disk to a compact disk with read only memory (CD-ROM) or a stored file on a Web server.

Delivery of new media products can take many forms. The most common form is on CD-ROM or floppy disk. With the use of a modem, users can also download software and multimedia packages from an online service and be billed for the transaction through a credit card or an account.

Digital media must be delivered to be used.

Interactive media puts you in the driver's seat

Most multimedia products are interactive. That is they use some sort of action or reaction from the user to prompt them to continue, by making objects available on the screen to "click" or select. Multimedia products provide what is called "nonlinear" access—the user can move around from place to place in the program by clicking on objects that link them to other parts of that program. This is unlike traditional ways where the user is typically expected to read or listen to the work from start to finish—passively.

Countless applications are being found for the use of this new technology. This can range from professional, business, educational, home computer, and some consumer markets.

The uses for digital media are endless. It is limited only by the imagination of its developers.

What law applies to multimedia?

There is no single governmental body overseeing the new media industry. Instead it is comprised of existing laws that are applicable to the development, protection, and distribution of the product. These include copyright law, (probably the biggest source of protection and concern), trademark and unfair competition law, laws protecting one's right of publicity and privacy, defamation law, trade secret laws, and patent laws. These are all of great concern to developers and designers. Or at least they should be.

Copyright law

All multimedia products, including all of the content generated, are automatically protected by federal copyright from the instant of their creation. This law creates a valuable form of protection for multimedia works that you personally produce. This law also acts as a double-edged sword in that it provides an obstacle for content that is created by others, in the event that your creation is similar to another or uses parts of another. Originality is the key.

Some information produced by others is public domain, free to use without permission from the creator. However, this part of the copyright law is interpreted by the federal law in different ways. The law itself can create problems from the very start because some of it does not seem to make sense.

Trademark and unfair competition law

Let us say that you are creating an educational program to show kids how to do fractions. You thought it would be a great idea to include a few of the Disney characters so the children could relate. Right off the bat a problem has arisen. These images are trademarked and copyrighted by the originator. The originator has the right to allow or disallow you the authority to use their images in your program.

There is the issue of names that can also hold trademarks. The name of the McDonald's cannot be used without the consent of the owner. Unfortunately, perfectly common words and phrases get trademarked and you are then restricted from using them. Related to the trademark law is the law regarding unfair competition. Say I have written a song and used certain sampling from another artist's song to finish out my piece. This would fall under the Unfair Competition Law because people could link me as being affiliated with the performer that I took the sampling from.

Right of publicity and privacy

When you produce the digital media disk or Web site of the Mickey Mouse Club, the creator of that good old rodent may not wish him to be a part of the production. This is their right under the Publicity Law—to negate the use of Mickey in your production.

If in the production of your new media package you use the names of people, living or dead, those people have the right to refuse you the use of their name under the Privacy Law. Even deceased people have rights, believe it or not.

Defamation law

If your program or project says something really nasty about Barney the Dinosaur, that he likes to eat Boy Scouts for lunch, you could get sued for that remark under Defamation Law. It could be seen as injuring the reputation of a large stuffed object and making false statements about purple dinosaurs. Honest.

Contract law—guilds and unions

If you used an outside company to produce the graphics used in a digital media production and this company operated using union talent, the union contract may have reserved the right to collect certain payments as a condition of using the content in a market other than for which it was originally developed. In some cases this may include payment to many different unions.

Copyright law

All software used to create a particular new media product, unless you created your own programs, including authoring and editing tools, is copyrightable. Any controlling software that is needed to run the program falls under the same copyright laws. Most programs have a "player" or "viewer" that are specifically given away so that productions can be played or viewed.

Trade secret law

In the creation of your multimedia product you may have needed to use valuable information from another company that wishes this information be kept a secret. Under the Trade Secret Law you must not pass on any facts that would harm the company that you borrowed the information from.

Like copyright laws, trade secret laws are automatically obtained. But you must not let the secret out or you lost them.

Patent law

Because of a law passed in November of 1993, patent law became an important consideration for multimedia creators. If you developed a process of navigating through a program and it is uniquely different from everyone else's then you can get a patent on it so no one else can use it without your permission. It must be unique and never used before.

Components of multimedia

For a complete understanding of the legal issues that surround the digital media industry we need to look at what areas in particular affect the components of digital media. This will undoubtably depend on whether you are dealing with the content or the technology of digital media.

Developing a digital media product involves creativity, collection of images and elements, adaptation of objects, combination of elements, coordination of resources, and synchronization of multiple and varying types of content from sound to animation. These specifically include text, images, color palettes, databases, characters, musical works, sound recordings, pictures, animations, still images, and motion pictures.

Digital media content

Virtually all content of a new media work is qualified for copyright protection. This includes text and databases, musical works, sound recordings, pictorial and graphic works, motion pictures, and other audio, visual, or textual works. Text by itself is protected by copyright. This includes items such as books, plays, poems, articles, editorials, as well as other works expressed in text form. Both fiction and nonfiction books are obviously copyrightable as well.

Databases as a whole are copyrightable. This copyright holds true if the database is comprised of text, numbers, or images. This database could hold images for a certain client of a company and without their permission, the images would not be usable by outside groups. An example of database that may not be copyrightable is one comprised of names and addresses, but a compilation copyright might be granted for the order or structure of the database.

Fictitious characters that exist in visual form such as Batman and Daffy Duck usually carry a copyright on them for their use in works. Usually these characters are copyrightable in image or pictorial form, and not so as much as in text form. Assume all cartoon characters are copyrighted.

Musical works

Any kind of musical work—verbal, written lyrics, or the music itself—is copyrightable. Lyrics by themselves do not sufficiently qualify for a copyright, but with the combination of sound and tone they do qualify.

To use music you may require rights from several different parties. The rights necessary depend on whether or not you record the music from its source or use the performance of a third party. Since the music will frequently be sound tracks from a particular TV show or motion picture, you will need to clear the rights to particular performances. Rights in music are quite complicated, so be very careful.

Mechanical rights include basic right to use a musical composition and do not include the right to publicly perform the music. A mechanical license often does not permit the use of the music with still or moving images—such use requires a "synchronization" license. Copyright law provides a compulsory license for mechanical rights, but most licensees prefer to obtain these rights commercially through the Harry Fox Agency or other agencies. This preference is based on the payment and accounting requirements imposed by the "compulsory" license in the Copyright Act.

If the music is to be synchronized with still or moving images on a screen, the licensee must obtain a "synchronization" license. Although these rights may also be handled by the Harry Fox Agency, in some cases you may need to contact the music publisher directly.

You will also need a license for public performance because your multimedia work will be shown to an audience—a public performance that is "open to the public" or at any place where a number of persons gather. Most music publishers permit either ASCAP or BMI to license their public performance rights. These rights do not apply to a particular performance by a particular individual or group to use the particular recording of a performance of the musical composition. Obtaining a mechanical license to "Feelings" would not permit you to use a particular artist's performance of the song. You can see how particular this all is.

If you have rights to a particular performance or recording and you desire the musical composition to be performed by a particular group or individual, you must also obtain the right of the copyright holder for that particular performance. The licenses described previously are limited solely to the right to use the musical composition. Unless you are prepared to have new artists record the music, you must negotiate with the holder of the rights. These rights are generally held by record companies or private companies.

Sound recordings that incorporate music, the spoken word, and other sounds, such as sound effects, are copyrightable regardless of the medium with which the recording is done on. Examples of sound recordings are spoken words, recordings of sounds such as trolley bells, thunder, bird calls, or virtually any other sound that can be recorded. You could have a case where two sounds sound alike but were both recorded as original sounds. Keep good records.

Photographs and still images

Whether drawn graphics or photos of art, they all are copyrightable. This can include two and three dimensional works of art, photographs, and prints of art reproductions. Illustrations such as architectural and technical drawings are also copyrightable. A picture of a statue is protectable.

Copyrights for photographs are owned by the photographer, although they may either be assigned to another party or transferred to the photographer's employer under the work-for-hire contract. The determination of who owns the appropriate rights in the photograph can be very difficult and time consuming because of fragmentation in this industry. For example, the fact that a photograph appeared in the *Forbes* does not necessarily mean that the *Forbes* owns the copyright in the photograph. *Forbes* may only have a license to use it once in its magazine. Common limitations in the licensing of photographs include the color of reproduction, the medium (newspapers, magazines, etc.), and attribution as well as those relating to numbers of copies.

The rights required for an interactive multimedia work would be quite different from those which are normally granted to use photographs. For example, the photograph may appear several times throughout the work and the number of its appearances could be controlled by the viewer. Such flexibility is quite different from the rights traditionally granted in photography contracts.

Audio and visual works

It is clearly stated in the Copyright Act of 1976 that all motion pictures and other related audio and visual works are copyrightable. Video is a related audio and visual work. You must distinguish between video which you have created and video for which you need to obtain rights. The "authors" of a videotape may include the actors, directors, scriptwriters, music composers and the camera people. To avoid the problems of joint ownership of copyright, you should obtain the appropriate agreements from all individuals involved.

The use of the short videoclips from a particular event may require multiple clearances—including clearing the music used in the clip, obtaining a license from the copyright owner, paying reuse fees to any entertainment unions such as SAG or the Directors Guild, and clearing the rights of publicity of the participants. If you use the "scripted" performances, you will have to pay reuse fees to the writers if they are members of the Writers Guild or even the actors.

Feature films

The use of feature films can be especially complex and expensive. Feature films are frequently based on a literary novel whose use is licensed to a studio. The film may also use music developed by a third party. The owner of the copyright in the film may not have the necessary rights to the music or the underlying *novel* to permit their use in the multimedia work. This situation is further complicated by provisions of the various motion picture industry guild agreements (such as the Screen Actors Guild or the Directors Guild of America) which require payment of fees upon incorporation of portions of a film or video into another work. You may also have to obtain rights of publicity, including releases from the individual actors depending on their contract with the studio.

Stock footage doesn't make it free

Stock footage is available from *stock houses* in many cities and ranges from historical footage of various locations or events to local commercials. Other institutions, such as television stations, may also license their newscasts. These institutions generally base their royalty on the type of use of the film or video. Different royalties are established for use on national television or local television broadcasts.

New media technology

The creation of a new media production is very complicated. Whether it be for navigating or searching your disk the software as well as the content is copyrightable. Copyright protects both content and software that go into the work, plus there is a copyright on the entire package itself. One multimedia work could have copyrights that belong to different parties. Since parts of a multimedia production use someone else's art or design, the copyrights can become quite numerous since these items are used solely with permission.

What rights do copyright owners have?

The Copyright Act of 1976 grants the owner of the copyright of a work a limited monopoly of the use of the work. To accomplish this, the copyright owner is given the right to do the following:

1. Make copies of the work yourself for whatever purpose (reproduction right)

2. Prepare derivative works (adaptation right)

3. Distribute copies of the work publicly by sale, rental, lease, or lending (distribution right)

4. Perform or display the work publicly in front of an audience (diaplay and public performance right).

When someone other than the copyright holder performs any of the above they are in direct infringement of the copyright law unless given the permission of the copyright owner.

Owners' reproduction right

The Copyright Act gives you, the copyright owner, an exclusive right to reproduce and to authorize the work to be reproduced or used by others. Because the owner of the copyright is the only person granted the right to authorize reproduction of a copyrighted work, the process of digitally scanning or capturing a copyrighted photograph from a magazine and then storing it onto a disk constitutes making a copy, and also constitutes copyright infringement. Copying a sound recording to a disk without authorization by the copyright owner also constitutes copyright infringement.

Content—your own or get permission

You have three choices with respect to the creation of multimedia work:

1. You can create your own text and graphics.

2. In certain cases you can use existing content without the permission of the author (fair use of content).

3. Or you can obtain permission to use preexisting content from the owners of the rights of that content which you want to use.

Relationship of copyright to digital media

Copyright protection is the most common and easiest form of legal protection that can be used for the written or digital media world. But, like most things, it has its positive and negative points. It is a great protector of work done by yourself, but it also works against you when you want to use someone else's ideas or content, since everything that is produced, in any medium, from the day of its inception, carries a federal copyright. For you to take full advantage of this legal issue you need to know what you can and can't get away with and what is and is not protected under a federal copyright.

What works are protected by copyright?

In reference to the Copyright Act of 1976, virtually everything that goes into a digital media production is copyrightable. Period. To fully understand the extent of copyright on digital or multimedia productions, we need to understand three areas: copyright protection for content (for example, text and music); copyright protection for the technology (software for writing programs); and the completed digital media product itself. They synthesize into one protectable unit.

Works protected

Copyright protection is available for all "works of authorship." This includes the following types of works which are of interest to the multimedia developer:

- Literary (text based) works.

Novels, nonfiction prose, poetry, newspaper articles, magazine articles, journal articles, computer software, software manuals, training manuals, technical manuals, catalogs, brochures, ads (text), and compilations such as business lists and directories.

- Musical (music based) works.

Songs, operas, musicals, ad jingles, and instrumentals.

- Dramatic works.

Plays, operas, and skits.

- Pictorial and sculptural works.

Photographs, posters, maps, paintings, drawings, graphic art, display ads, cartoon strips, cartoon characters, stuffed toys, statues, paintings, and fine art.

- Motion pictures and other visual works.

Movies, documentaries, travelogues, training videos, television shows, television ads, and interactive multimedia works.

- Sound (sound based) recordings.

Recordings of music, sounds, or words.

Copyright protection

Copyright protection is automatic when an "original" work of authorship is "fixed in a tangible medium of expression." Registration with the Copyright Office is optional (but you have to register before you file an infringement suit.) Registering early may make you eligible to receive attorney's fees and statutory damages in a future lawsuit.

A work is original as far as copyright is concerned if it is original to the author and was not copied from some preexisting work or works.

A work is "fixed" when it is made "sufficiently permanent or stable to permit it to be perceived, reproduced, or otherwise communicated for a period of more than transitory duration." Even copying a computer program into RAM has been found to be of sufficient duration for it to be "fixed" (although some legal scholars disagree with this conclusion).

Neither the "originality" requirement nor the "fixation" requirement is stringent. An author can "fix" words, for example, by writing them down, typing them on a typewriter, dictating them into a tape recorder, or entering them into a computer. A work can be original without being novel or unique—or even correct.

Only minimal creativity is required to meet the originality requirement. No artistic merit or beauty is required. A work can incorporate preexisting material and still be original. When preexisting material is incorporated into a new work, the copyright on the new work covers only the original material contributed by the author.

A developer's digital media work incorporates a number of photographs that were made by a photographer (who gave the developer permission to use the photographs in the multimedia work). The digital media work as a whole owes its origin to the developer, but the photographs do not. The copyright on the digital media work does not cover the photographs, just the material created by the developer.

Scope of protection

Copyright protects against *copying* the *expression* in a work as opposed to the idea of the work. The difference between *idea* and *expression* is one of the most difficult concepts in copyright law. The most important point to understand is that the protection of the *expression* is not limited to exact copying—either of the literal words of a novel or the shape of a stuffed bear. Copyright infringement extends to new works which are *substantially similar.* A copyright owner has five exclusive rights in the copyrighted work:

Reproduction Right: The right to reproduce, duplicate, transcribe, or imitate the work in fixed form.

Modification Right: The right (also known as the derivative works right) is the right to modify the work to create a new work. A new work that is based on a preexisting work is known as a "derivative work."

Distribution Right: The distribution right is the right to distribute copies of the work to the public.

Public Performance Right: The right to show the work at public place or to transmit it to the public. In the case of a motion picture or other audio or visual work, showing the work's images in sequence is "performance." Some types of works, such as sound recordings, do not have a public performance right.

Public Display Right: The public display right is the right to show a copy of the work directly or by means of a film, slide, or television image at a public place or to transmit it to the public. In the case of a motion picture or other audiovisual work, showing the work's images out of sequence is considered "display."

In addition, certain types of works of "visual art" also have "moral rights" which limit the modification of the work and the use of the author's name without permission from the original author.Anyone who violates any of the exclusive rights of a copyright owner is an infringer.

An artist scanned a photographer's copyrighted photograph, altered the image by using digital editing software, and included the altered version of the photograph in a multimedia work that artist sold to consumers. If the artist used the photograph without permission they infringed the copyright by violating the reproduction right (scanning the photograph), the modification right (altering the photograph), and the distribution right (selling the altered photograph in their work). Three counts and you're out!

A copyright owner can recover actual or, in some cases, statutory damages (which can be as high as $100,000 in some cases) from an infringer. In addition, courts have the power to issue injunctions (legal orders) to prevent or restrain copyright infringement and to order the impoundment and destruction of infringing copies.

The extent (term) of copyright protection depends on three major factors:

1. Who created the work
2. When it was created
3. When it was first distributed commercially

For copyrightable works created on and after January 1, 1978, the copyright term for those created by individuals is the life of the author plus 50 years. The copyright term for *work made for hire* is 75 years from the date of first *publication* (distribution of copies to the general public) or 100 years from the date of creation, whichever expires first.

The copyright is owned by the person (or persons) who create the work. If the work is created by an employee within the scope of their employment, the employer owns the copyright because it is a *work for hire.* The copyright law also includes another form of *work for hire* that applies only to certain types of works which are specially commissioned works. These include audio and visual works, which will include most interactive media projects. In order to qualify the work as a *specially commissioned* work for hire, the creator must sign a written agreement stating that it is a *work for hire* prior to commencing development of the product. Most foreign jurisdictions do not recognize the *specially commissioned* work for hire, and you need an assignment to transfer rights in some countries. Do not assume foreign law is the same as U.S. law.

Avoiding copyright infringement

Current technology makes it too easy to combine material created by others—audio and video segments, music, graphics, photographs, text—into a multimedia product. The technical ease of copying does not give you the legal right to do so. If you use copyrighted material owned by others without getting permission you can incur legal liability for hundreds of thousands or even millions of dollars in damages.

Most of the third-party material you will want to use in a multimedia product is protected by copyright. Using copyrighted material without getting permission—either by obtaining an "assignment" or a "license"—can have disastrous legal consequences. The owner of the copyright can prevent the distribution of your product (an injunction) and obtain damages for infringement, even if you did not intentionally include their material. An assignment is generally understood to transfer all of the intellectual property rights in a particular work, although an assignment can be more limited in scope. A license provides the right to use a work and is generally quite limited in its scope.

No copyright notice is required on the product

Most published works present a copyright notice. However, for works published on or after March 1, 1989, the use of copyright notice is *optional*. The fact that a work does not have a copyright notice doesn't mean that the work is not protected. This is the most ridiculous thing we have ever heard, but it is the law.

How much copying is allowable? None!

It is correct that *de minimis* copying (copying a *small* amount) is not copyright infringement. Unfortunately, it is rarely possible to tell where *de minimis* copying ends and copyright infringement begins. How small is small?

Copying even a small amount of a copyrighted work is infringement if that which is copied is a qualitatively substantial portion of the copied work. As example, a magazine article that used 300 words from a 300,000-word autobiography was found to infringe the copyright on the autobiography. Even though the copied material was only a small part of the autobiography, the copied portions were among the most important passages in the autobiography.

Copying any part of a copyrighted work is fraught with danger. If what you copy is truly a tiny and nonmemorable part of the work, you may get away with it (the work's owner may not be able to tell that your work incorporates an excerpt from the owner's work). However, you run the risk of having to defend your use in expensive litigation. If you are copying, it is better to get a permission or a license (unless fair use applies). You will escape liability for infringement by demonstrating how much of the protected work you did not take.

If you give credit to the author, you are not a plagiarist because you are not pretending that you authored the copied work. However, attribution is not a defense in cases of copyright infringement.

Assume nothing

Don't assume that a copyright owner will be happy to have you use their work. Even if the owner is willing to let you use the work, the owner will probably want a license fee. Content owners view *new* media as a *new* market for licensing their material. In 1993, freelance writers sued the *New York Times* and other publishers over the unauthorized publication of their work through online computer services. And music publishers have sued CompuServe, a commercial computer service, over the distribution of their music online.

You cannot escape liability for copyright infringement by altering or modifying the work you copy. If you copy and modify elements of a copyrighted work, you will be infringing the copyright owner's modification right as well as the copying right. Remember, ideas and facts are not protectable, but the expression of those ideas or facts is.

Fair use is fairly complex

You don't need a license to use a copyrighted work if:

- Your use is fair use.
- The work you use is in the public domain.
- The material you use is factual or an idea.

It is often difficult to tell whether a particular use of a work is fair or unfair. Determinations are made on a case-by-case basis by considering these factors:

1. The courts are most likely to find fair use where the use is noncommercial, such as book review or school project.

2. The courts are most likely to find fair use where the copied work is a factual work rather than a creative one.

3. The courts are likely to find fair use where a tiny amount of the protected work is used. If small in amount, but substantial in terms of importance, you could be in trouble.

4. The courts are most likely to find fair use where the new work is not a substitute for the copyrighted work.

5. The courts are likely to find fair use if the new work does not compete with it, either in whole or part

If your new media work serves traditional *fair use* purposes—criticism, comment, news reporting, teaching, scholarship, research—you are within the bounds of fair use more than if your work is sold for commercial gain.

Public domain

Public domain works—works not protected by copyright—can be used by anyone. Because these works are not protected by copyright, no one can then claim the exclusive rights of copyright for such works. The plays of Shakespeare are in the public domain; there never was a copyright. Works enter the public domain because the term of the copyright expired, the copyright owner failed to "renew" their copyright under the old Copyright Act of 1909, or the copyright owner failed to properly use a copyright notice—for works created before March 1, 1989. After that copyright notice became optional.

Ideas or facts are fair game

You don't need a license to use facts or ideas from a protected work. The copyright does not extend to the work's facts or ideas. Copyright protection is limited to original works of authorship, and no one can claim originality or authorship of facts or ideas. You don't need a copyright license for material which you create yourself.

The rules regarding ownership of copyright are complex. You should not assume that you own the copyright if you pay an independent contractor to create the work (or part of it). The copyright for a work is owned by the individual who creates the work, except for full-time employees working within the scope of their employment and copyrights which are assigned in writing.

The pertinence of patent law

This law protects inventions and processes ("utility" patents) and ornamental designs ("design" patents). Inventions and processes protected by utility patents can be electrical, mechanical, or chemical in nature. Works protected by design patents are a design for a lamp, a design for silverware, and a design for a machine. Typefaces are not patentable unless linked to special hardware, such as machine readable fonts. There have been notable lawsuits, even one suing the Patent Office, but typeface designs are not protectable. Their names *can* be trademarked.

A utility patent covers an invention that must be new, useful, and "nonobvious." To meet the "new" requirement, the invention must not have been known or used by others in the United States before the applicant invented it, and it also must not have been patented or described in a printed publication in the U.S. or a foreign country before the applicant invented it. The policy behind the novelty requirement is that a patent is issued in exchange for the inventor's disclosure to the public of the details of their invention. If the inventor's work is not novel, the inventor is not adding to the public knowledge, so the inventor should not be granted a patent. At least that is the underlying philosophy.

To meet the nonobvious requirement, the invention must be sufficiently different from existing technology and knowledge so that, at the time the invention was made, the invention as a whole would not have been obvious to a person having ordinary skill in that field. The policy behind this requirement is that patents should only be granted for real advances, not for mere technical tinkering or modifications of existing inventions.

Utility patents are granted for a period of 17 years. Design patents are granted for a period of 14 years. Once the patent on an invention or design has expired, anyone is free to make, use, or sell the invention or design.

Trademark law

Trademarks and service marks are words, names, symbols, or devices used by manufacturers, marketers, and service providers to identify their goods and services, and to distinguish their goods and services from goods manufactured and marketed by others. For trademarks used in commerce, federal trademark protection is available under the federal trademark statute, the Lanham Act. Many states have trademark registration statutes that resemble the Lanham Act, and all states protect unregistered trademarks under the common law (nonstatutory law) of trademarks.

Trademark protection is available for words, names, symbols, or logos that are capable of identifying or distinguishing the owner's goods or services from the goods or services of others. A trademark that merely describes a class of goods rather than distinguishing the trademark owner's goods from goods provided by others is not protectable.

"Corn flakes" is not protectable as a trademark because it describes a type of cereal that is sold by a number of companies rather than distinguishing one brand from another. It is said to be "generic."

A trademark that is so close to a trademark already in use in the U.S. which might cause confusion is not protectable. In addition, trademarks that are "descriptive" of the functions, quality or character of the goods or services have special requirements before they will be protected and these are quite complex.

The most effective trademark protection is obtained by filing a trademark registration application in the Patent and Trademark Office. Federal law also protects unregistered trademarks, but is limited to the geographic area in which the mark is actually being used. State trademark protection under common law is obtained simply by adopting a trademark and using it in connection with goods or services. This protection is limited to the geographic area in which the trademark is actually being used. State statutory protection is obtained by filing an application with the state trademark office.

Rights of publicity, libel and other laws

There are several other areas of law that deal with the right of the individual to control their image and reputation. Libel and slander laws protect an individual against the dissemination of falsehoods. To be actionable, the falsehood must injure his or her reputation or subject them to hatred, contempt or ridicule. The individual can obtain monetary losses as well as damages for mental anguish.

So there you have it

Copyright infringement will pose a problem with electronic documents. In the future, with so many documents being transferred and people having accessibility to various documents and images with the ability to cut and paste, it will not be easy to protect copyrights and copyright laws.

So, be original. Create your own content. Borrow nothing. Marry a lawyer.

We repeat: This chapter is not offering nor does it claim to offer legal advice. Consult an attorney, especially one who specializes in copyright law. The material presented is derived from a number of sources and attempts to offer an outline of current copyright practices. It is neither complete nor comprehensive. And lastly, do not assume that any of it is right. Only Courts determine that.

10.

TELECOMMUNICATIONS

The communications bill of 1996 will turn the telephone business and the entire telecommunications industry upside down. The overhaul of telecommunications law will be good and bad. In the end the results should be lower prices all around and a headlong rush to offer advanced communication services over fiber optic networks and satellites. The breakup of AT&T was a headache for consumers years ago but then long distance telephone prices plunged nearly 70%. And just as the phone companies plan to offer cable service, cable operators are gearing up to provide local phone service. Everyone wants a piece of the action.

The result is an industry driven almost entirely by the forces of business and technology, and its impact is almost certain to be jarring. The new bill obliterates monopolies in local telephone service and cable television by letting the industries invade each other's markets. Cable television prices will be freed from regulation. The cost of basic local telephone service may creep up; long distance prices are likely to decline. Confusion will abound, but so will new services for consumers.

Television, radio and newspaper publishing may also be transformed. We could see an increase in the number of television and radio stations and newspapers a single company can own, paving the way for a large-scale consolidation of mom-and-pop media organizations into nationwide chains. Local telephone companies would be forced to open their markets to new rivals. They would immediately be allowed to enter the cable television business and eventually the long distance market as well.

Fewer than 1% of all cable systems face a significant

competitor, and cable systems with exclusive franchises charged about 30% more, on average, than those that did have competitors. A local company would have to let rivals interconnect with its own network so that competing companies would be able to link calls between each other's customers. Customers, for the first time, would be allowed to keep their telephone numbers if they switch to a different carrier. New direct-to-home broadcast satellites, like Hughes' Communications DirecTV, now offer more 150 channels of movies and television programming, including all major cable networks. For now, though, competition in cable seems more theoretical than real.

Direct-broadcast-satellite providers like DirecTV and Prime Star are signing up customers rapidly. But direct-broadcast satellite still has fewer than 2 million subscribers—compared with 61 million for cable. But with the muscle of AT&T and MCI, some analysts estimate that the industry could have 10 million customers by 2000.

Small cities and towns will not see competition for years. Bell Atlantic and NYNEX are merging and hope to offer cable programming over a relatively inexpensive microwave technology known as "wireless cable."

MCI Communications, the nation's second-largest long-distance carrier, has announced a $2 billion plan to build local networks in major cities. Sprint Corp., the third-largest long-distance carrier, has teamed up with four of the country's biggest cable television companies to offer a full range of wired and wireless services at the local level.

Under the bargain, negotiated to permit transition to a new system of subsidies, 10 states will be able to require dialing parity to use a rival to the local company by dialing a "1" instead of a five-digit code to reach a rival carrier.

Since the Baby Bells can now compete in the long-distance market, they will focus on the AT&T Corp. and the MCI Communications Corp., rather than on cable giants like Time Warner Inc. or Tele-Communications Inc. Long distance is a much bigger business than cable, and the Bells already own much of the equipment needed to provide that service.

Technology, ho!

Making a dramatic leap in the speed of communications, researchers have succeeded in transmitting information at a rate of 1 trillion bits a second through an optical fiber. This is like transmitting the contents of 300 years' worth of daily newspapers in a single second, or 12 million telephone conversations simultaneously through a wire made of glass.

The experiments were carried out by Fujitsu Ltd., the Nippon Telegraph and Telephone Corp., and a team from AT&T Research and Lucent Technologies, both of which were formerly parts of Bell Laboratories. Reaching a transmission of one trillion bits, or a terabit was not expected for years. Optical fibers convey information as a stream of pulses of light representing the ones and zeros, or bits, of computer code. The fibers are used for telephone communications, computer data transmissions, and video transmissions. The more capacity a fiber has, in general, the lower the cost of communications. Higher capacities are needed to carry video signals, for digital media, including video-on-demand.

Computer manufacturers have been striving to reach speeds of 1 trillion operations a second, known as a teraflops. The results achieved by the three groups are two and a half times as fast as the previous experimental record of 400 billion bits a second. It is 400 times faster than the fastest commercial systems today, which carry 2.5 billion bits, or 2.5 gigabits, a second. Systems with capacities of 10 gigabits a second and 20 gigabits a second are out now.

All three approaches have one thing in common. Instead of sending one stream of light through the fiber, as is done today, they sent multiple streams of light, each a slightly different wavelength, thereby multiplying the amount of information that can be transmitted. A potential drawback is that the new technique is likely to be used for relatively short distances, perhaps 50 miles. Light signals degrade and have to be amplified or reconstituted en route and with multiple light streams this becomes very expensive.

Wireless

Wireless communication is absolutely exploding. Americans spent roughly $22 billion on wireless services in 1995, and is expected to double by the turn of the century. Along with Sprint and two other big cable companies eager to break into the local phone business, Cox has invested $2.1 billion for licenses to offer a new cellular service, called personal communications services, or PCS. The partners, which include Tele-Communications Inc. and Comcast, are spending $2 billion more to build a national network capable of reaching 144 million people. Rather than waging a ground war for each other's wired network, many companies are taking their battles to the skies—trying to infiltrate the well-protected markets of their rivals by beaming telephone or television signals directly to customers.

Long distance carriers are particularly keen on catching the air wave, because they have national networks but lack direct connections to individual telephone and cable households. AT&T announced that it would invest $137.5 million in a satellite broadcasting service, DirecTV. A week later MCI paid $682 million for an orbital slot to offer its own high-powered satellite broadcasting service. Sprint and its cable partners reorganized their alliance to focus on wireless, rather than wired services—even changing the venture's name to Sprint Spectrum, an allusion to the radio-frequency spectrum used by wireless services.

The popularity of wireless stems partly from the fact that the business has already grown so rapidly. The cellular industry that made McCaw so rich went from almost nothing in 1980 to a $20 billion business last year. And in the last few years, refinements in digital technology have also made wireless transmission suitable for a broader range of services, including computer-precise television transmissions and high-speed data transfer. Even after the telecommunications market is fully deregulated, the high cost of stringing wires or laying cables, and the expected resistance of the local cable and phone-wire monopolies to interlopers, will make it tough for long-distance carriers, local telephone companies, or cable operators to break into each other's traditional landline markets.

Just as the cable industry has been reluctant to take on

the Bell companies, the Baby Bells have been reluctant to go after the cable industry. Several of the Bells announced plans years ago to upgrade their copper wire networks to carry video services. By plunging into the satellite broadcasting business, however, AT&T and MCI will eventually be able to beam 150 channels of programming directly from satellites to pizza-pan-sized dishes on rooftops.

The Baby Bells own the wires that link into most American households. Would-be rivals like long distance carriers can either negotiate to lease capacity on the local network, try to duplicate the network or find ways to bypass it. To duplicate that kind of a network, they must dig up streets or put up telephone poles. So far cellular has been too expensive to become a viable competitor to local phone service, but personal communications services may have a better shot.

PCS networks, using digital radio transmissions for voice and data communications via small hand-held devices, can serve many more customers simultaneously than current cellular networks. That fact, and the cut-throat competition as several wireless companies compete in many of the larger metropolitan markets, should result in prices much lower than cellular companies have typically offered.

The business of telecommunications

Electronic information services grew 16% in 1995 to an estimated $16.6 billion, slightly faster than the previous year's 15% increase. Most of the industry's revenues come from business users of information for financial management, research, marketing, purchasing, and general business administration. On-line services had 3.8 million business subscribers and 4.4 million consumer subscribers. Consumer and small business use of on-line services for news, information, transactions, and bulletin boards account for an increasing share of the industry's growth.

About 65% of information is delivered on-line via computers equipped with modems, but the amount of information delivered on CD-ROM continues to grow, reaching 18% in 1995. Delivery by telephone generated about $690 million in revenue in 1995, even though government regulations have curtailed growth. Audiotext is especially popular for delivery, in real time, of stock quotes, mortgages rates, sports and lottery results, weather, and news. A major newspaper publisher estimated that more than one million calls will be made monthly to its audiotext service.

The commercialization of the Internet has also enhanced awareness of electronic information services and encouraged the development of new services. Progress was made in 1995 on interactive multimedia, a technology that combines computer, telephone, and television services. This technology could attract more at-home users to information services.

The U.S. industry derives about 35% of its revenue from foreign sources. Nearly all major U.S. suppliers of electronic information services are established overseas. The European Community (EC) and Japan are the largest overseas markets. The EC wants to develop a fully integrated electronic market for information services, but it has been hampered by legal, administrative, language, and cultural barriers. France, with nearly six million users of the Minitel on-line service, is the largest European provider of electronic information.

U.S. companies are strong overseas because of the widespread use of the English language and some of the most advanced and diversified services.

ISDN solutions

The digital protocols that allow ISDN lines to carry voice and data faster than the speediest modems have been around for nearly a decade, but users are still labeled as early adopters for installing ISDN. The most common ISDN line, Basic Rate Interface (BRI), has two 64-Kbps channels for data transmission rates of up to 128 Kbps. That translates to file transfers of about 1 Mbyte a minute, or two to four times the speed of a V.34 modem.

That speed can be improved by file compression and by the new crop of PCI adapter cards for Power Macs, which can bundle as many as four ISDN lines for a total of eight 64-Kbps transmission channels.

While ISDN, or Integrated Services Digital Network, has spread in Europe—first in pre-press and then in other industry segments—in the United States, the regional Bell operating companies have been slower in updating switches with ISDN to provide nationwide service coverage. Phone companies have lagged at pumping up technical support to aid users with ISDN installations.

Compounding the difficulty for earlier adopters of the technology was a lack of interoperability. While service providers are supposed to follow the ISDN specification promulgated by the International Telecommunications Union, not all have adhered to the letter. In California, for example, Pacific Bell's slight deviation from the specification means its BRI ISDN lines run at 56 Kbps per channel rather than the normal 64 Kbps. Despite the slow start, analysts see a bright future for ISDN. About 250,000 ISDN installations will be completed by the end of 1996, escalating to 1 million lines by the end of the century.

In the publishing world, ISDN's biggest draw is its ability to replace overnight courier services and SyQuest cartridges with a faster file delivery mechanism. In a July 1995 survey by British Telecom Ltd. of 200 United Kingdom-based ISDN sites, more than 60% of respondents reported using their lines an average of four times per day. These users were sending files that averaged between 5 and 10 Mbytes. The mix of advertising agencies, reproduction and design bureaus responding to the survey said their most common use of ISDN was to send files to printers, followed by sending files to clients for proofing and approval.

Strong demand doesn't remove installation stumbling blocks and operating problems, but there are strong signs that U.S. regional Bell operating companies realize ISDN's potential and are redoubling coverage and support efforts. While coverage varies from vendor to vendor, on average about 70% of the U.S. population should be able to get an ISDN line by the end of 1996, according to figures from Dataquest Inc.

It might also be easier for converts to survive the ISDN installation process, now that the phone companies and equipment vendors have agreed on a set of ordering codes that make it easier for installers to pinpoint the correct line interface for each customer's equipment and application.

Smaller companies without a telecommunications manager or consulting budget may have the rockiest installation road. After potential users find out if service is available in their area and learn about some of the installation pitfalls, the question is which type of connection to consider: point-to-point links based on adapter cards or a WAN using ISDN-enabled routers.

Forming a WAN with ISDN has its own set of advantages, including the ability to offer file transfers to the entire network and the ability to multiplex lines together for greater aggregate bandwidth. On the downside, however, routers must use some of the ISDN pipe for standard network messaging, and in AppleTalk's case that can lead to considerable bandwidth loss. Below are the percentages of each company's service area that will be ISDN-capable.

	End of 1994	**End of 1996**
Ameritech Corp.	78	82
Bell Atlantic Corp.	72	82
Bell South Corp.	70	72
GTE Personal Comm. Services	36	54
Nynex Corp.	46	53
Pacific Bell	80	86
Southwestern Bell Corp.	60	76
U.S. West Inc.	47	53

Source: Dataquest Inc.

Conclusion

We are entering the new age of communications. Eventually it will be possible to tap into a network from anywhere to anywhere, via wire or wireless. We will all be connected.

11.

THE FUTURE

When we think of media we often return to the metaphor of the book. It is a common symbol for information delivery. Its form metamorphosed into magazines and newspapers and catalogs and directories, but the simple book is under attack from new media.

The book is by far the most popular form of reading available to all people. The Koreans invented the bound book and Johannes Gutenberg gave us the technology to mass produce them. Where will book publishing go? How will it get there? What will be the fate the book as it makes the transition into the 21st century?

Some would say that they imagine a book publishing facility as a place where thousands of the same titles are printed for mass distribution. This is not completely true. Only a small percentage of the books printed every year are popular enough to require a mass production. It is usually those on the best selling lists that make it to a high production run. There has been a steady decline on the run length of books, even those on the best selling lists. This is due to the success of the personal computer and the technological advancements that go along with it. A main factor as to why the computer is causing book production to decrease is the use of multimedia and user interactivity of the computer. Technical manuals which need diagrams can now be displayed on the computer with a compact disk as the medium. A compact disk can also store video and sound files as well.

The book publisher of the future will be printing shorter runs of books because of the success of the computer. The book in no way will ever be replaced by the computer however, people will still feel more comfortable with the document in their hands rather than reading it on a screen.

The biggest change that the book publisher will see is the handling of the files. We are living in an electronic world where transferring information over phone lines can now be done with ease. With this technology the book publisher can now have "digital" customers anywhere in the world.

The file transfer rates are always increasing which cuts down on the amount of time needed to send the entire book electronically. There are several options the book publisher has as far as outputting is concerned. One technique is to output directly to film from the computer. There are imposition software programs available to help with the proper page layout so that it is printed correctly with the right amount of signatures needed. Then the films can be used to make plates which will eventually go on the press. As long as there are no revisions with that particular book the plates can be used over again to make more copies. Making changes in the book is not really a problem as long as the electronic file is saved. Another way to output is to go directly to the plate or to the press. In the latter case a blank plate is mounted on the plate cylinder on the press for each unit and then a laser images the information right onto the plate. It is estimated that a four color process job can go from the image on the monitor of the workstation to four process plates on the press in 3 minutes.

The publisher may produce fewer books; however, the demand for on demand printing will increase. Printers may print more books on an as needed basis rather than just mass produce them. Also, the use of electronic file transferring will be greatly used to quicken the printing process.

The book publisher of the future will be doing the same process as in the past. The tools, equipment, and methods will change, but in the end we will still have a nicely bound book to read and enjoy for use again and again.

But let's talk about the bookstore of the future. Books are handled 10 to 20 times from press to buyer. Short runs will be the norm. Why not move the book production to the buyer? Bookstores in the future could have digital printing systems. Books could be selected through connection to the Copyright Clearance Center and extracted from one of many archives. The file would be downloaded to the bookstore and printed out and bound.

Publisher and retailer get paid. The buyer gets a book on demand. True, the cost of the equipment and the speed of its operation must increase. It may be that the bookstore works in conjunction with a quick copy center.

Library of the future

The principal content of a library is usually printed or written material in the form of books or periodicals, and the principal means of information transfer is reading. Many libraries include collection of films, microforms, phonograph records, sound tapes, etc., which transfer information by means other than reading. The whole purpose of a library is to provide an access to information.

Libraries of the future will deal with access to electronic documents. They will contain not only electronic or linear (mainly printed) documents, but also a hybrid of the two. In this sense a library would act like a publisher-cum-book seller. These libraries could thus provide printing of electronic texts or graphics onto paper, on demand. There will still be archival books and printed matter, but mostly there will be databases and electronic files, for viewing or printing.

A fact that cannot be overlooked is that the information from the so-called *Electronic Libraries* of the future may be used purely in its digital form. We must keep in mind while evaluating the future of libraries the changing circumstances of the library users. Many of the users are themselves adopting electronic habits, making the use of telecommunication, computers and electronic storage, and local printout more popular. One can safely assume that this trend would continue for the years to come. But, they will still print out pages for whatever reason and libraries will find that they are as much in the printing business as many printers.

This leads us to the importance of telecommunications in the field of library services. Access to a paper library is limited to the people who physically visit the library. This limitation restricts the number of users of a particular library and hence the information base. Electronic data coupled with efficient and cheap telecommunications leads to extremely fast access of information by people situated even a long distance from the library. The reader does not need to know where the disk drives are physically located so long as the records appear on the screen. This opens up the possibility of having only one "universal" library in place of thousands that we have today. The material can be shared by users anywhere from their workstations, be it at their place of work or their their home. The library, hence does not only have to be accessed during any specific *library hours.* This is easier in terms of maintainability as well as accessibility, provided the necessary equipment is readily available. *Education Technology News* said that Pacific Bell has offered a wiring system so that all public libraries in the California service territory will establish a public-access network and will offer electronic services and information which will be freely accessible for the public. Bell decided to wire the libraries with Integrated Services of Digital Network (ISDN) and apply low cost computers and video conferencing equipment.

The form of information

The form in which the information is obtained or presented is as important as the information itself. The presentation of documents in the case of paper libraries is governed primarily by the authors and the publishers of the books. On the contrary, electronic documents usually have the flexibility of being viewed as per the requirements of the user. As computer technology advances in the terms of more sophisticated software and hardware, the user has more and more control over the search and access procedures. The convenience of a book still cannot be underestimated.

Printed pages are excellent in terms of the resolution they provide to the eye. They offer a convenient format which is easily movable and does not require any use of electronic equipment to read (except spectacles for some).

Electronic documents with or without the generation of paper copies, become preferable:

1. When dealing with highly volatile documents.
2. When communication is required.
3. When searching information through documents.
4. When manipulating documents.
5. When browsing through documents.

Economic factors that govern the usage of electronic documents are very crucial. As of today, not many countries of the world have monetary strength to support convenient electronic libraries. Most people agree that the cost of electronic equipment can be expected to come down drastically, while expenses of labor increase. Also technological advances would increase the flexibility of the electronic library. But until the time this is possible, most of us will be connected with paper libraries. Laptop computers have already given substantial mobility to computers and hence to electronic documents. As the cost-effectiveness of electronic libraries increases, their popularity would also increase. If it outweighs the cost-effectiveness of the book or paper libraries, it is very difficult to ascertain if the paper library would really became obsolete or not. In my opinion, a hybrid with more emphasis on the electronic library, would exist in that case. Electronic libraries could create a distance between the rich and the poor. Due to less literacy the poor would not be able to afford this new technology.

The new papyrus

In 1985, a shiny silver disk filled with data and information called the CD-ROM was introduced. Today using this technology an entire database can be stored on a single disk. An entire book or an encyclopedia can be transferred to a CD-ROM in a digital format. An easy accessibility to desired information can be derived. One can also store video clips and sounds with related documents. The CD-ROM is the new papyrus, but its technology is changing too quickly. Libraries cannot afford too many kinds of equipment to read all the different media.

The library of the future would be in an electronic form, but the need for books will not wane. One would be able access a library from home.

The catalog of the future

Traditionally catalogs have provided us with a means of shopping without leaving home. That is because a new catalog system is on its way. The catalog of the future started as the CD-ROM catalog but seems destined for the Web.

Even though CD-ROM and the online shopping are still

in their infancy, there is every indication that they will revolutionize the way catalog shopping is done and it is most likely to be the preferred way of catalog shopping of the future. One of the advantages of shopping online is that it will cut down the time it usually takes to order something from a catalog as no mailing will be necessary. Paper catalogs can be expensive to print, mail and distribute. On the other hand, a CD-ROM which has the capacity of holding thousands of paper catalogs can cost as much as a first class stamp to mail. Again CD-ROMs are less expensive to manufacture. A 650 megabyte CD-ROM can cost as little as 70 cents to produce and can contain more information than a paper catalog. In addition, a paper catalog has short shelf life, and it is static and un-interactive.

Network shopping

You can now shop in cyberspace—in an electronic shopping mall. You can shop in any part of the country or world you choose. Since there is no overhead, it is likely that the price of commodities will be cheaper than in stores. Shopping online for products such as cars or other major purchases can arm one with information about the product that can put them in a stronger bargaining position. Shoppers can also download price lists and other information on products they want to buy. When the system is perfected, searching for products will be easy since one can narrow their search by different categories such as brand name and by using search and retrieve commands.

Grocery shopping

Another service that has been added to the online shopping is grocery shopping in some big cities such as San Francisco. One can also buy airline tickets, rent a car or a movie. Blockbluster has begun experimenting with online movie rentals. Many companies plan to create what they call *smart agents* that will collect detailed information from consumers and target products specifically at those individuals. AT&T PersonaLink Services also promises a mall-like interface that will allow users to browse through a graphical storefront containing aisles of products. Merchants will also be able to monitor customers buying habits and then offer a discount or some other help when a customer has walked into online for threes week and buys nothing. Some of the new technology promises a moving demonstration with sound which allows customers to see how a product can be used before they make a purchase. Customers that will benefit most may be music buyers. Play the screen sample and hear the songs before you buy the CDs.

Digital shopping

Since many families have both cable and computers, some companies are exploring ways to merge the two together. Intel is developing an adapter that could be used to hook both computer and television together. Plans are also underway at making the television more like a computer. This will make it possible for people to shop interactively from their television. But it does not appear that interactive television will be out in the next five to ten years. When this technology is available, one would be able to call up products of interest any time of the night or day. It is a scary thought—24 hour shopping.

If online shopping has this many advantages to offer why aren't we all shopping online? The technology is still in its infancy. With the exception of CD-ROM catalogs being used to distribute software there is not really much going on in commercial catalog production now. However some mail order retailers are hitting the information highway but still offering very few items compared to what they offer in their regular catalogs. JC Penny, Spiegel, Lands and Hammacher Schlemmer offer electronic versions of their catalogs online. Pictures are rare and of poor quality. It appears that the best things to buy online at the moment are those things that one does not need to see before buying. Entries are vague and one can search for hours and end up getting frustrated.

Electronic catalogs

Despite its drawbacks it appears that the electronic catalog has much promise because it has shown to have a lot of capabilities that can be improved on, which has made it a worthwhile product to pursue. It will change the way business is conducted. However its drawbacks will keep paper catalog in use almost indefinitely even if in small quantity. Also the mall will still keep flourishing until this technology is perfected. Many people like to stroll the malls and see and try things they want to buy like clothes and shoes so mall traffic will continue flowing no matter how convenient it becomes to shop online. Plus, one must consider the social aspects of the mall. It's not the same on the computer.

Magazine publishing

Magazine publishers were limited in the ways they could communicate articles, stories and pictures. They were limited to print. Over the years the use of magazines exploded as printing technology increased to provide longer runs and more colors. In time, quality improved and more magazines were produced in larger numbers covering thousands of different topics of concern. Eventually, the digital age arrived on the commercial publishing scene. In this environment text and photographs were able to be manipulated via the computer. And all was good.

Advancements in technology have led us to new methods of information dissemination. We are no longer limited to print. The future of magazine publishing is already here. The future will consist of vast improvements of existing systems. Almost all information produced today is created in digital form. This digital form allows for endless possibilities in magazine publishing. Speed, quality, and quantity now know there are no limitations. The key here is that content is not only going to go to press—it can go anywhere, in any form.

Tailored publications

Publications will be even more tailored to the individual. Besides the common ink jet address and name, databases will become truly customized. Some see a future where nearly every piece of printed material you receive is tailored to your taste. If you always order clothes from L.L. Bean in blue and green, why waste postage and catalog space promoting red items? With more powerful and efficient ways to create and store databases of personal information and buyer behavior, why should a company waste time with things they know you will not buy? Magazine publishing in the future will be so customized that even junk mail will look good.

The receivers of magazines have become easily and obviously identifiable. With such customization in a narrower direct market, people will receive magazines based on a remote interest. Therefore, we will see the creation of even more magazines, as people begin to realize that others across the world have the same interest as they. Start-up magazines will begin to enter the market as it becomes easier for people to publish themselves.

Presses will also be faster than ever. Press speeds are now reaching well over 2,000 feet per minute. Speed is only one factor in the advancements of presses. Gapless printing will play a big role in increased production, helping to reduce waste in makeready, by adding more area to print, and paving the way for automatic blanket changing. A press design that eliminates an allowance on the cylinder circumference for a blanket-mounting gap permits both a larger live print area and a smaller trim allowance which means less waste and savings in paper.

Computer to Plate (CTP) will play an important role in magazine publishing because time between pre-press and press is greatly reduced and last minute changes can be made much easier. This will drastically improve turnaround time. Technology similar to that used on the Heidelberg GTO-DI will be adapted for larger-sized and longer run web presses.

Together this will eliminate the need for creating plates off press. Time will also give way to the end of film. With the declining price and increasing quality and storage of digital cameras, the need for film will become obsolete. Of course it will take time to wean us off of film. Efficient digital proofing systems must also be developed if we truly desire to eventually replace film.

Other great pressroom improvements will drastically change the publication of magazines. With digital information being transferred, a CTP system can also do pre-sets for ink, speed, water, compensation and registration. A system like this will not only benefit magazine publishing but will also help and change the face of the press crew of the future.

The press will be controlled from a control room, not from the console on the floor. This could be a factor in reducing costs of labor and injuries. Plates mounting themselves, blankets and rollers washing themselves, and paper automatically loaded and unloaded is the destiny of the press room.

In other words, print is not dead or even dying. Publishers will be able to get any run at high quality at a lower cost. Printed magazines will continue.

Moving the information

Digital information is already being transferred via dedicated phone lines to distribute information. Companies like R.R. Donnelley, Banta Corp. and many others rely on this to increase turnaround time and keep the entire process virtually in-house. Magazine and newspaper publishers are not limited to just one site for production. The use and speed of phone lines will be drastically increased. This will make a tremendous difference for the client, publisher, and advertiser to create and work on files in *real time.*

With these advancements industry is trying to stay one step ahead by creating special protocols and quicker means to send information. Switched Multi-megabit Data Services (SMDS) can be integrated with existing ISDN technologies. SMDS is a standardized connectionless, high speed, public packet-switched data service.

Non traditional publications

Digital files and digital technologies are now enabling us to do a better job handling workflow in the print shop, but that's not all. We have begun to enter the true future of magazine publishing not with the press, but with Compact Disk Read Only Memory and the World Wide Web. With the small yet growing number of people using computers and exploring the wonderful uses of CD-ROM and the WWW, magazine publishing has and will continue to venture into the uses of digital media technology.

Disk publishing

CD-ROM publishing will greatly expand its use as a means for digital publishing. CDs will be used to find and retrieve, copy and paste information. With the use of faster speed CD-ROM drives and with a reduction in prices for CD writers, not only will we see an increase in do-it-yourself publishers, but use will increase among commercial publishers.

Space becomes an important factor. A stack of 12 magazines compared to that of fewer than 12 CDs of a years subscription plus supplements, is not only accommodating for space, but truly makes magazines a treasured reference set that is easy to use and access.

Cyber pubs

Some new publications are not on your local newsstand—they're in cyberspace. They are published online or on CD-ROM. Publishers with good ideas are skipping the costs, headaches and risks of tackling the nation's overcrowded newsstands and heading directly into cyberspace. Their products for now are tagged with the watered-down labels *e-zines* or *digizines*, but they clearly represent an entirely new medium.

Magazines began the rush to cyberspace more than two years ago. Commercial online services, led by America Online, recognized that magazine titles such as *Time* and *Car and Driver* came with built-in audiences and reputations. In effect, these magazines became trusted safe havens for consumers new to the online world. But the vast opportunities opened up by the growth of new media platforms—combined with some publishers' hesitancy to explore electronic editions—has opened the door for a swarm of pioneers.

Magazines distributed through commercial online services or the Internet are accessible to wide audiences but still face the problem of luring them in. CD-ROM titles are trying to move beyond the shelves of computer software stores to gain a toehold on newsstands and in bookstores, music stores and other specialty retailers. Bundling with computer hardware or traditional magazines is another option.

Offbeat marketers have quickly embraced offbeat digizines. The second issue of pop culture disk *Blender*, for example, lists eight advertisers. The marketers include alternative magazine *Paper* and the New York Web, an Internet service.

Many print magazines haven't taken full advantage of

those strengths. They looked at their early forays into electronic publishing as a way to digitize and *repurpose* existing editorial material and are only now recognizing the need to create entirely new products.

The future is now

Digital media—print and non-print— are advancing. The publisher of tomorrow is not just big companies. The publisher of tomorrow is you and me. From our computer we command a host of new approaches to information dissemination. *We are the press.*

12.

AFTERWORD

Where will you work in the future? Based on information and telecommunications technology being developed today you can probably work anywhere you want. All you need is a telecommunications connection and a computer—plus the knowledge to use digital media.

Almost every job and career will be touched by digital media. Doctors will review medical tests from thousands of miles away; librarians will reference books in libraries in another part of the world; you will order clothing from a store in Australia from the comfort of your home; and you will have access to almost any piece of information anywhere on the globe. At work you will confer with others long distance on large video screens that let you show charts and videos of the information you have found so that workers can interface from any location.

All of this is possible because of the specialists who make digital media work for you and me. In the past, digital technology described telephone workers who wired buildings or established telephone service. Today a telecommunications specialist must be a whiz at computers with knowledge of the electronic links between them. Those links include normal phone lines as well as cellular and satellite communications. The last two eliminate wires, although the future will involve both wire and wireless approaches.

One kind of wire, called fiber optics cable, allows many more signals to travel by using pulses of light rather than electrical charges. New wire and wireless technology allow more data to flow which will permit sound and video to be handled as routinely as text.

The computer and you

In 1980 the desktop personal computer was introduced

and life has not been the same since. Early versions required operators to type complicated codes to make the computer perform some task, like add numbers or write a memo. The computer is not the tool of tomorrow; it is the tool of today.

Careers in computer technology include specialists in hardware and software. The hardware is the computer and all of its mechanical and electronic parts; the software is the programming that lets the hardware perform specific tasks, like adding a column of numbers, or editing a memo, or drawing a picture. Most computer users do not have to be hardware and software specialists—they only need to know how to use certain programs to do their job. Some computer specialists create the basic programs that perform specific tasks, while most computer users apply those programs in the course of their work.

Linking computers into networks

From the beginning computers could be connected over telephone lines. These devices changed computer signals to telephone signals so the data could travel over telephone wires to another computer where another modem changed the signal back into computer form. Next, computers were inter-connected in offices and buildings with special wires and software called *local area networks.* Computer users could now transfer files and share information without leaving their desks. The *LANs* as they are called were then connected over telephone lines so that other LANs could connect and pretty soon businesses could connect with their customers and customers could connect with their suppliers. This resulted in *WANs* or Wide Area Networks that linked wider and wider groups of computer users.

The Internet is the network of networks

In the 1960s the U.S. Government began to develop a special network that could connect the military in a manner that was redundant—this meant that if one part of the network was out of use, another part could take over. This network of networks eventually grew to encompass other government workers and universities and research scientists. It is called the *Internet* and today it has millions of connections all over the world, including many high schools, businesses and industries. But the Internet was hard to use because it required computer coding and most of its use was for *e-mail*—messages typed on one computer and sent to another computer. Then came the WWW.

The World Wide Web

In the early 1990s one part of the Internet that was used by scientists advanced the idea of linking information, not just computers. Research information was encoded to contain *hyperlinks,* words or phrases that the user could click on, and then a related piece of information would instantly appear on the computer screen. That new piece of information could come from a computer in another part of the globe.

Information is usually stored in *databases,* which organize and categorize the data so that a user can search and retrieve. A database can contain text and sound and images.

Resourceful users developed *browsers* so that these links could contain more than text. Users could hear sounds and see pictures, graphics, or even videos. Browsers eliminated the users' need to type complex codes. You just had to point and click. This pointing and clicking lets you jump from information link to information link, from site to site. Hyperlinking lets you go directly to a related piece of information or a related location. Users then "surf the net" as they navigate from one location to another, no matter where it is.

Cyberspace and the virtual world

The world of electronic information is also called *cyberspace.* It is not the real world but rather an artificial version of it. The term *virtual* is used to describe something that is not quite real and cyberspace is also called *virtual reality.* Web sites are virtual stores or businesses and eventually we will shop in a virtual world. Many kinds of businesses are already on the WWW and their home pages display pictures of their products with descriptions and ordering information.

Career paths

To make all this work an army of specialists is required to create the information, establish the links, layout the look of the screens, and maintain the network. The reason the system is easy to use is that talented people created and produced it with that goal in mind. You may point and click, but someone must create the program that makes the right thing happen when you want it to happen.

The entire entity is now a *system*, a combination of hardware, software, networks, databases, and people all organized to find, acquire, process, and output information for decision making purposes. In this new Information Age, new careers are being developed that require a blend of computer and telecommunications skills so that users will be able to find and use information easier and faster.

The digital world

You can visit *Thomas* at the Library of Congress and read the *Congressional Record* instead of waiting for it to be printed and delivered. Click and jump to CNN and read the latest reports and see video clips of latebreaking news. Click again and visit the Disney home page and see clips of current films or schedules for theme parks. Check out classified job listings, or read the latest on cyberspace from *HotWired*. Search through the works of Shakespeare and download one of the hundreds of free texts in *Project Gutenberg*. Listen to live radio; see live TV or hold a telephone conversation, to say nothing of typing letters to anyone and e-mailing them without stamps. Leave comments with NBC news and check out a political candidate. Find financial information on public corporations through *Edgar* at the SEC and review the latest Ford automobiles. Download software updates or get answers to computer problems. Get new product information instantly and then find a source through the electronic *Yellow Pages*. Read magazines and newsletters. Visit virtual stores and review their merchandise. It's all there in the world of digital media waiting for you to point and click.

Gutenberg didn't know what he started. The first revolution in moveable type changed the way people stored, accessed and transmitted knowledge. The second revolution does the same. Here we go again.

INDEX

(Adobe) Acrobat — 50, 52, 54, 56, 57, 60, 68, 71, 72, 127
(Adobe) Illustrator — 64
3M — 26
addPS — 72
America Online — 38, 121, 133
AppleLink — 133
Archie — 123
ARPANET — 119
ASCII — 49, 63, 65, 76
AT&T — 17, 35, 164, 165, 166, 167, 175
ATM — 56, 59, 67
Berners-Lee, Tim — 122
Bolt, Beranek & Newman — 122
book — 7, 169, 170, 171
bookstore — 8
Canon Color Laser Copier — 28
CD-DA (Compact Disc — Digital Audio) — 138
CD-I — 41, 140
CD-ROM XA — 139, 140
CD-ROM — 7, 11, 12, 15, 16, 21, 23, 32, 35, 36, 37, 40, 41, 42, 49, 54, 127, 128, 129, 137, 138, 139, 140, 141, 142, 144, 168, 173, 174, 176, 179, 181, 182, 183
cellular telephone — 10
CERN (European Laboratory for Particle Physics) — 124, 125
CGATS (Committee for Graphic Arts Technologies Standards) — 60
Chromapress — 34
CompuServe — 38, 121, 133
Computer Bulletin Board System — 37
Congressional Record — 23
Copyright Act — 151
copyright law — 146
copyright — 153
cordless telephone — 10
cybermedia — 41

cyberspace — 36, 187
Cycolor — 27
DAT — 21
database — 35, 187
DDAP (Digital Distribution of Advertising for Publications) — 62
Delphi — 38
digital ad — 60
digital press — 25
digital printing — 31
DVD (digital video disk) — 41, 137, 142
e-mail — 7, 8, 11, 20, 22, 54, 120
EPS — 70
(Farallon) Replica — 52, 54
GEnie — 38, 133
Gibson, William — 18
GIF — 110
Gopher — 123
GTO-DI — 28
hot link — 105, 106
Howtek — 28
HTML (Hypertext Markup Language) — 75–118, 125, 180
HTTP (Hypertext Transport Protocol) — 79, 80, 88, 89
hypermedia — 41, 43
hypertext — 63
Indigo — 29
intellectual property — 143
interactive media — 43, 46
Internet — 119, 120, 122, 133
ISDN (Integrated Services Digital Network) — 129, 130, 172
ISO (International Standard Organization) — 87
JPEG — 59, 69, 80, 110
Kodak — 27
Library of Congress — 188
library — 171, 172
LZW — 67, 69
(Macromind) FreeHand — 64
magazine — 7, 33
MCI — 121, 164, 167
modem — 10
MPEG — 81, 87, 117
multimedia — 43, 44, 143, 147, 152
(NCSA) Mosaic — 78, 81
NCSA (National Center for Supercomputing Applications — 78

Netscape — 81, 109, 126, 127
newspaper — 7
(No Hands) Common Ground — 52, 54, 55, 58, 59
NYNEX — 164, 167
patent law — 160
PC with CD-ROM — 10
PC — 10
PCMCI — 41
PDF — 53, 55, 57, 61, 62, 63, 64, 65, 66, 67, 68, 69, 70, 71, 72
Pixelmaster — 28
portable document — 49
PPP — 81, 134
Prodigy — 133
public domain — 159
QuarkXPress — 52, 63, 71
Quickmaster-DI — 28
QuickTime — 87
RTF — 76
satellite — 9
SGML (Standard Generalized Markup Language) — 75, 76, 84
SLIP — 81, 134
TCP/IP — 119, 120, 125, 134, 136
Tele-Communications — 166
telecommunications — 163
telephone — 10
TIFF/IT — 61
trademark law — 160
UNIX — 15, 88, 133, 134
URL — 79, 106, 107, 109
VCR — 10
WAIS (Wide-Area information Servers) — 123
WAN (Wide Area Network) — 131
Windows NT — 15
WordPerfect Envoy — 52
WORM (Write Once Read Many) — 140
WWW (World Wide Web) — 14, 16, 41, 45, 78, 79, 80, 81, 82, 83, 88, 92, 122, 129, 179, 187
Xeikon — 29
Xerox — 26, 28

AUTHOR

Frank. J. Romano is the *Melbert B. Cary, Jr. Distinguished Professor of Graphic Arts, College of Imaging Arts and Sciences, Rochester Institute of Technology.* With a career that has spanned over 30 years in the printing and publishing industries, he is the author of eleven books, co-editor of the International Paper *Pocket Pal,* contributor to major encyclopedias and dictionaries, and the author of numerous articles.

He has founded eight publications, serving as publisher or editor or both. He lectures extensively throughout the United States and Canada, having addressed virtually every club, association, group and professional organization at one time or another and consulted for major corporations, publishers, government and other users of digital media.

He has been quoted in the *New York Times, Wall Street Journal, Times* of London, and many other newspapers and publications, as well as on TV and radio. Professor Romano teaches digital prepress, electronic publishing, and digital printing.